ASE Test Preparation

Automobile Certification Series

Service Consultant (C1)

5th Edition

Australia • Brazil • Japan • Korea • Mexico • Singapore • Spain • United Kingdom • United States

ASE Test Preparation: Automobile Certification Series, Service Consultant (C1), 5th Edition

Vice President, Technology and Trades Professional Business Unit: Gregory L. Clayton

Director, Professional Transportation Industry Training Solutions: Kristen L. Davis

Editorial Assistant: Danielle Filippone

Director of Marketing: Beth A. Lutz

Marketing Manager: Jennifer Barbic

Senior Production Director: Wendy Troeger

Production Manager: Sherondra Thedford

Content Project Management: PreMediaGlobal

Senior Art Director: Benjamin Gleeksman

Section Opener Image: Image Copyright baranq, 2012. Used under license from Shutterstock.com

ISBN-13: 978-1-111-12712-1

ISBN-10: 1-111-12712-3

Delmar Cengage Learning
5 Maxwell Drive
Clifton Park, NY 12065-2919
USA

Cengage Learning is a leading provider of customized learning solutions with office locations around the globe, including Singapore, the United Kingdom, Australia, Mexico, Brazil, and Japan. Locate your local office at: **international.cengage.com/region**.

Cengage Learning products are represented in Canada by Nelson Education, Ltd.

For more information on transportation titles available from Delmar, Cengage Learning, please visit our website at **www.trainingbay.cengage.com**.

For more learning solutions, please visit our corporate website at **www.cengage.com**.

Printed in the United States of America
1 2 3 4 5 6 7 16 15 14 13 12 11

Table of Contents

Preface

Delmar, a part of Cengage Learning, is very pleased that you have chosen to use our ASE Test Preparation Guide to help prepare yourself for the Automobile Service Consultant (C1) ASE certification examination. This guide is designed to help prepare you for your actual exam by providing you with an overview and introduction of the testing process, introducing you to the task list for the Automobile Service Consultant (C1) certification exam, giving you an understanding of what knowledge and skills you are expected to have in order to successfully perform the duties associated with each task area, and providing you with several preparation exams designed to emulate the live exam content in hopes of assessing your overall exam readiness.

If you have a basic working knowledge of the discipline you are testing for, you will find this book is an excellent guide, helping you understand the "must know" items needed to successfully pass the ASE certification exam. This manual is not a textbook. Its objective is to prepare the individual who has the existing requisite experience and knowledge to attempt the challenge of the ASE certification process. This guide cannot replace the hands-on experience and theoretical knowledge required by ASE to master the vehicle repair technology associated with this exam. If you are unable to understand more than a few of the preparation questions and their corresponding explanations in this book, it could be that you require either more shop-floor experience or further study.

This book begins by providing an overview of, and introduction to, the testing process. This section outlines what we recommend you do to prepare, what to expect on the actual test day, and overall methodologies for your success. This section is followed by a detailed overview of the ASE task list to include explanations of the knowledge and skills you must possess to successfully answer questions related to each particular task. After the task list, we provide six sample preparation exams for you to use as a means of evaluating areas of understanding, as well as areas requiring improvement in order to successfully pass the ASE exam. Delmar is the first and only test preparation organization to provide so many unique preparation exams. We enhanced our guides to include this support as a means of providing you with the best preparation product available. Section 6 of this guide includes the answer keys for each preparation exam, along with the answer explanations for each question. Each answer explanation also contains a reference back to the related task or tasks that it assesses. This will provide you with a quick and easy method for referring back to the task list whenever needed. The last section of this book contains blank answer sheet forms you can use as you attempt each preparation exam, along with a glossary of terms.

OUR COMMITMENT TO EXCELLENCE

Thank you for choosing Delmar, Cengage Learning for your ASE test preparation needs. All of the writers, editors, and Delmar staff have worked very hard to make this test preparation guide second to none. We feel confident that you will find this guide easy to use and extremely beneficial as you prepare for your actual ASE exam.

Delmar, Cengage Learning has sought out the best subject-matter experts in the country to help with the development of *ASE Test Preparation: Automobile Certification Series, Automobile Service*

Consultant (C1), 5th Edition. Preparation questions are authored and then reviewed by a group of certified, subject-matter experts to ensure the highest level of quality and validity to our product.

If you have any questions concerning this guide or any guide in this series, please visit us on the web at **http://www.trainingbay.cengage.com**.

For web-based online test preparation for ASE certifications; please visit us on the web at **http://www.techniciantestprep.com/ to learn more**.

ABOUT THE AUTHOR

Jerry Clemons has been around cars, trucks, equipment and machinery throughout his whole life. Being raised on a large farm in central Kentucky provided him with an opportunity to complete mechanical repair procedures from an early age. Jerry earned an Associate in Applied Science degree in Automotive Technology from Southern Illinois University and a Bachelor of Science degree in Vocational, Industrial and Technical Education from Western Kentucky University. Jerry has also completed a Master of Science degree in Safety, Security and Emergency Management from Eastern Kentucky University. Jerry has been employed at Elizabethtown Community and Technical College since 1999 and is currently an Associate Professor for the Automotive and Diesel Technology Programs. Jerry holds the following ASE certifications: Master Medium/Heavy Truck Technician, Master Automotive Technician, Advanced Engine Performance (L1), Truck Equipment Electrical Installation (E2), and Automotive Service Consultant (C1). Jerry is a member of the Mobile Air Conditioning Society (MACS), as well as a member of the North American Council of Automotive Teachers (NACAT). Jerry has been involved in developing transportation material for Delmar, Cengage Learning for seven years.

ABOUT THE SERIES ADVISOR

Mike Swaim has been an Automotive Technology Instructor at North Idaho College, Coeur d'Alene, Idaho since 1978. He is an Automotive Service Excellence (ASE) Certified Master Technician since 1974 and holds a Lifetime Certification from the Mobile Air Conditioning Society. He served as Series Advisor to all nine of the 2011 Automobile/Light Truck Certification Tests (A Series) of Delmar, Cengage Learning ASE Test Preparation titles, and is the author of *ASE Test Preparation: Automobile Certification Series, Undercar Specialist Designation (X1), 5th Edition*.

SECTION 1

The History and Purpose of ASE

ASE began as the National Institute for Automotive Service Excellence (NIASE). It was founded as a non-profit, independent entity in 1972 by a group of industry leaders with the single goal of providing a means for consumers to distinguish between incompetent and competent technicians. It accomplishes this goal through the testing and certification of repair and service professionals. Though it is still known as the National Institute for Automotive Service Excellence, it is now called "ASE" for short.

Today, ASE offers more than 40 certification exams in automotive, medium/heavy duty truck, collision repair and refinish, school bus, transit bus, parts specialist, automobile service consultant, and other industry-related areas. At this time there are more than 385,000 professionals nationwide with current ASE certifications. These professionals are employed by new car and truck dealerships, independent repair facilities, fleets, service stations, franchised service facilities, and more.

ASE's certification exams are industry-driven and cover practically every on-highway vehicle service segment. The exams are designed to stress the knowledge of job-related skills. Certification consists of passing at least one exam and documenting two years of relevant work experience. To maintain certification, those with ASE credentials must be re-tested every five years.

While ASE certifications are a targeted means of acknowledging the skills and abilities of an individual technician or service consultant, ASE also has a program designed to provide recognition for highly qualified repair, support and parts businesses. The Blue Seal of Excellence Recognition Program allows businesses to showcase their technicians, service consultants, and their commitment to excellence. One of the requirements of becoming Blue Seal recognized is that the facility must have a minimum of 75 percent of their technicians ASE certified. Additional criteria apply, and program details can be found on the ASE website.

ASE recognized that educational programs serving the service and repair industry also needed a way to be recognized as having the faculty, facilities, and equipment necessary to provide a quality education to students wanting to become service professionals. Through the combined efforts of ASE, industry, and education leaders, the non-profit organization entitled the National Automotive Technicians Education Foundation (NATEF) was created in 1983 to evaluate and recognize academic programs. Today more than 2,000 educational programs are NATEF certified.

For additional information about ASE, NATEF or any of their programs, the following contact information can be used:

National Institute for Automotive Service Excellence (ASE)
101 Blue Seal Drive S.E.
Suite 101
Leesburg, VA 20175
Telephone: 703-669-6600
Fax: 703-669-6123
Website: **www.ase.com**

SECTION 2

Overview and Introduction

Participating in the National Institute for Automotive Service Excellence (ASE) voluntary certification program provides you with the opportunity to demonstrate you are a qualified and skilled professional technician who has the "know-how" required to successfully work on today's modern vehicles.

EXAM ADMINISTRATION

Through 2011, there are two methods available to you when taking an ASE certification exam:

- Paper and pencil
- Computer Based Testing (CBT)

Note: Beginning 2012, ASE will no longer offer paper and pencil certification exams. They will offer and support CBT testing exclusively.

Paper and Pencil Exams

ASE paper and pencil exams are administered twice annually, once in the spring and once again in the fall. The paper and pencil exams are administered at over 750 exam sites in local communities across the nation.

Each test participant is given a booklet containing questions with charts and diagrams where required. All instructions are printed on the exam materials and should be followed carefully. You can mark in this exam booklet but no information entered in the booklet is scored. You will record your answers using a separate answer sheet. You will need to mark your answers using a #2 pencil only. Upon completion of your exam, the answer sheets are electronically scanned and the answers are tabulated.

Note: Paper and pencil exams will no longer be offered by ASE after 2011. ASE will be converting to a completely exclusive CBT testing methodology at that time.

CBT Exams

ASE also provides CBT exams, which are administered twice annually, once in the winter and once again in the summer. The CBT exams are administered at test centers across the nation. The exam content is the same for both the paper and pencil and CBT testing methods.

If you are considering the CBT exams, it is recommended that you go to the ASE website at ***http://www.ase.com*** and review the conditions and requirements for this type of exam. There is also an exam demonstration page that allows you to personally experience how this type of exam operates before you register.

Effective 2012, ASE will only offer CBT testing. At that time, CBT exams will be available four times annually, for two-month windows, with a month of no testing in between each testing window:

- January/February – Winter CBT testing window
- April/May – Spring CBT testing window
- July/August – Summer CBT testing window
- October/November – Fall CBT testing window

Please note, testing windows and timing may change. It is recommended you go to the ASE website at ***http://www.ase.com*** and review the latest testing schedules.

UNDERSTANDING TEST QUESTION BASICS

ASE exam questions are written by service industry experts. Each question on an exam is created during an ASE-hosted "item-writing" workshop. During these workshops, expert service representatives from manufacturers (domestic and import), aftermarket parts and equipment manufacturers, working technicians, and technical educators gather to share ideas and convert them into actual exam questions. Each exam question written by these experts must then survive review by all members of the group. The questions are designed to address the practical application of repair and diagnosis knowledge and skills practiced by technicians in their day-to-day work.

After the item-writing workshop, all questions are pre-tested and quality-checked on a national sample of technicians. Those questions that meet ASE standards of quality and accuracy are included in the scored sections of the exams; the "rejects" are sent back to the drawing board or discarded altogether.

Depending on the topic of the certification exam, you will be asked between 40 and 80 multiple-choice questions. You can determine the approximated number of questions you can expect to be asked during the Service Consultant (C1) certification exam by reviewing the task list in Section 4 of this book. The five-year recertification exam will cover this same content; however, the number of questions for each content area of the recertification exam will be reduced by approximately one-half.

> ***Note:*** Exams may contain questions that are included for statistical research purposes only. Your answers to these questions will not affect your score, but since you do not know which ones they are, you should answer all questions in the exam.

Using multiple criteria, including cross-sections by age, race, and other background information, ASE is able to guarantee that exam questions do not include bias for or against any particular group. A question that shows bias toward any particular group is discarded.

TEST-TAKING STRATEGIES

Before beginning your exam, quickly look over the exam to determine the total number of questions that you will need to answer. Having this knowledge will help you manage your time throughout the exam to ensure you have enough available to answer all of the questions presented. Read through each question completely before marking your answer. Answer the questions in the order they appear on the exam. Leave the questions blank that you are not sure of and move on to the next question. You can return to those unanswered questions after you have finished the others. These questions may actually be easier to answer at a later time once your mind has had additional time to consider them on a subconscious level. In addition, you might find information in other questions that will help you recall the answers to some of them.

Multiple-choice exams are sometimes challenging because there are often several choices that may seem possible, or partially correct, and therefore it may be difficult to decide on the most appropriate answer choice. The best strategy, in this case, is to first determine the correct answer before looking at the answer options. If you see the answer you decided on, you should still be careful to examine the other answer options to make sure that none seems more correct than yours. If you do not know or are not sure of the answer, read each option very carefully and try to eliminate those options that you know are incorrect. That way, you can often arrive at the correct choice through a process of elimination.

If you have gone through the entire exam, and you still do not know the answer to some of the questions, *then guess*. Yes, guess. You then have at least a 25 percent chance of being correct. While your score is based on the number of questions answered correctly, any question left blank, or unanswered, is automatically scored as incorrect.

There is a lot of "folk" wisdom on the subject of test taking that you may hear about as you prepare for your ASE exam. For example, there are those who would advise you to avoid response options that use certain words such as *all, none, always, never, must,* and *only,* to name a few. This, they claim, is because nothing in life is exclusive. They would advise you to choose response options that use words that allow for some exception, such as *sometimes, frequently, rarely, often, usually, seldom,* and *normally.* They would also advise you to avoid the first and last option (A or D) because exam writers, they feel, are more comfortable if they put the correct answer in the middle (B or C) of the choices. Another recommendation often offered is to select the option that is either shorter or longer than the other three choices because it is more likely to be correct. Some would advise you to never change an answer since your first intuition is usually correct. Another area of "folk" wisdom focuses specifically on any repetitive patterns created by your question responses (e.g., A, B, C, A, B, C, A, B, C).

Many individuals may say that there are actual grains of truth in this "folk" wisdom, and whereas with some exams, this may prove true, it is not relevant in regard to the ASE certification exams. ASE validates all exam questions and test forms through a national sample of technicians, and only those questions and test forms that meet ASE standards of quality and accuracy are included in the scored sections of the exams. Any biased questions or patterns are discarded altogether, and therefore, it is highly unlikely you will experience any of this "folk" wisdom on an actual ASE exam.

PREPARING FOR THE EXAM

Delmar, Cengage Learning wants to make sure we are providing you with the most thorough preparation guide possible. To demonstrate this, we have included hundreds of preparation questions in this guide. These questions are designed to provide as many opportunities as possible to prepare you to successfully pass your ASE exam. The preparation approach we recommend and outline in this book is designed to help you build confidence in demonstrating what task area content you already know well while also outlining what areas you should review in more detail prior to the actual exam.

We recommend that your first step in the preparation process should be to thoroughly review Section 3 of this book. This section contains a description and explanation of the type of questions you will find on an ASE exam.

Once you understand how the questions will be presented, we then recommend that you thoroughly review Section 4 of this book. This section contains information that will help you establish an understanding of what the exam will be evaluating, and specifically, how many questions to expect in each specific task area.

As your third preparatory step, we recommend you complete your first preparation exam, located in Section 5 of this book. Answer one question at a time. After you answer each question, review the answer and question explanation information located in Section 6. This section will provide you with instant response feedback, allowing you to gauge your progress, one question at a time, throughout this first preparation exam. If after reading the question explanation you do not feel you understand the reasoning for the correct answer, go back and review the task list overview (Section 4) for the task that is related to that question. Included with each question explanation is a clear identifier of the task area that is being assessed (e.g., Task A.1). If at that point you still do not feel you have a solid understanding of the material, identify a good source of information on the topic, such as an educational course, textbook or other related source of topical learning, and do some additional studying.

After you have completed your first preparation exam and have reviewed your answers, you are ready to complete your next preparation exam. A total of six practice exams are available in Section 5 of this book. For your second preparation exam, we recommend that you answer the questions as if you were taking the actual exam. Do not use any reference material or allow any interruptions in order to get a feel for how you will do on the actual exam. Once you have answered all of the questions, grade your results using the answer key in Section 6. For every question that you gave an incorrect answer to, study the explanations to the answers and/or the overview of the related task areas. Try to determine the root cause for missing the question. The easiest thing to correct is learning the correct technical content. The hardest things to correct are behaviors that lead you to an incorrect conclusion. If you knew the information but still got the question incorrect, there is likely a test-taking behavior that will need to be corrected. An example of this would be reading too quickly and skipping over words that affect your reasoning. If you can identify what you did that caused you to answer the question incorrectly, you can eliminate that cause and improve your score.

Here are some basic guidelines to follow while preparing for the exam:

- Focus your studies on those areas you are weak in.
- Be honest with yourself when determining if you understand something.
- Study often but for short periods of time.
- Remove yourself from all distractions when studying.
- Keep in mind that the goal of studying is not just to pass the exam; the real goal is to learn.
- Prepare physically by getting a good night's rest before the exam, and eat meals that provide energy but do not cause discomfort.
- Arrive early to the exam site to avoid long waits as test candidates check in.
- Use all of the time available for your exams. If you finish early, spend the remaining time reviewing your answers.
- Do not leave any questions unanswered. If absolutely necessary, guess. All unanswered questions are automatically scored as incorrect.

Here are some items you will need to bring with you to the exam site:

- A valid government or school-issued photo ID
- Your test center admissions ticket
- Three or four sharpened #2 pencils and an eraser
- A watch (not all test sites have clocks)

> *Note:* Books, calculators, and other reference materials are not allowed in the exam room. The exceptions to this list are English-Foreign dictionaries or glossaries. All items will be inspected before and after testing.

WHAT TO EXPECT DURING THE EXAM

Paper and Pencil Exams

When taking a paper and pencil exam, you will be placing your answers on a sheet that requires you to blacken (bubble) in your answer choice.

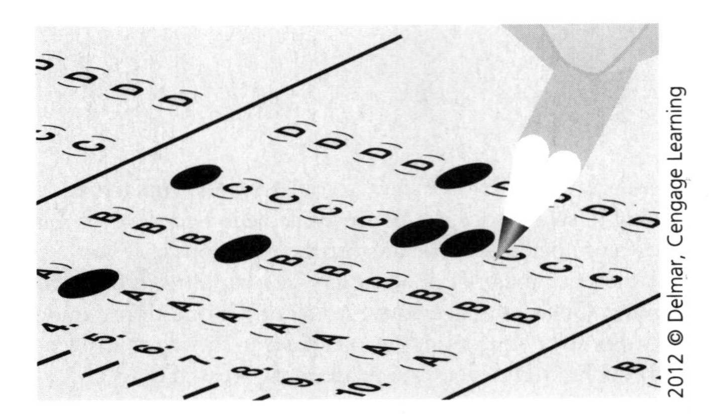

2012 © Delmar, Cengage Learning

Be careful that only your answers are visible on the answer sheet. Stray pencil marks or incomplete erasures may be picked up as an answer by the electronic reader and result in a question being scored incorrectly.

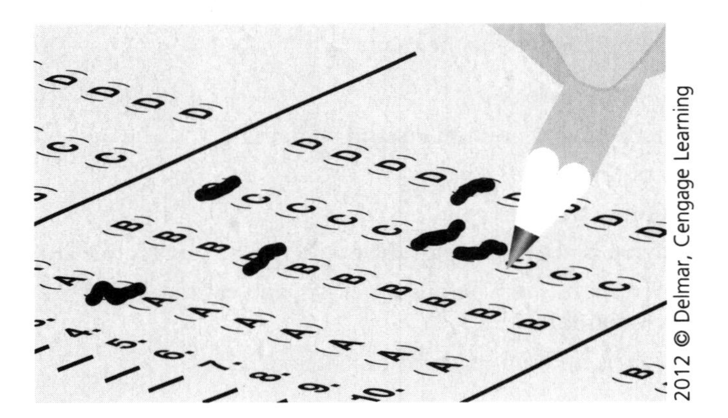

2012 © Delmar, Cengage Learning

Studies have shown that one of the biggest challenges an adult faces when taking a test that uses a bubble-style answer sheet is to place their answers in the correct location. To avoid problems in this area, be extra mindful of how and where you mark your answers. For example, when answering question 21, blacken the correct, corresponding bubble on the answer sheet for question 21. Pay special attention to this process when you decide to skip a question to come back to later. In this situation, many people forget to also leave the corresponding line on the bubble answer sheet blank as well. They inadvertently place their answer for the next question on the answer bubble sheet line that should have been left as a blank placeholder for the unanswered, skipped question. Providing a correct question response on the incorrect bubble answer sheet line will likely result in that question being marked wrong. Remember, the answer sheet for the paper and pencil exam is machine scored, and the machine can only "read" what you have blackened or bubbled in.

If you finish answering all of the questions on an exam and have time remaining, go back and review the answers for those questions that you were not sure of. You can often catch careless errors by using the remaining time to review your answers. Carefully check your answer sheet for blank answers or missing information.

At practically every exam, some technicians will invariably finish ahead of time and turn their papers in long before the final call. Since some technicians may be doing a recertification test and others may be taking fewer exams than you, do not let this distract or intimidate you.

It is not wise to use less than the total amount of time that you are allotted for an exam. If there are any doubts, take the time for review. Any product can usually be made better with some additional effort. An exam is no exception. It is not necessary to turn in your exam paper until you are told to do so.

CBT Exams

When taking a CBT exam, as soon as you are seated in the testing center you will be given a brief tutorial to acquaint you with the computer-delivered test, prior to taking your certification exam(s). Unlike paper and pencil testing, when taking a CBT exam, you will not have to worry about stray pencil marks or ensure that your answers are marked on the correct and corresponding answer bubble sheet line. The CBT exams allow you to select only one answer per question. You can also change your answers as many times as you like. When you select a second answer choice, the CBT will automatically unselect your first answer choice. If you want to skip a question to return to later, you can utilize the "flag" feature, which will allow you to quickly identify and review questions whenever you are ready. Prior to completing your exam, you will also be provided with an opportunity to review your answers and address any unanswered questions.

TESTING TIME

Paper and Pencil Exams

Each ASE paper and pencil exam session is four hours. You may register for and take anywhere from one to a maximum of four exams during any one exam session. It is recommended, however, that you do not register for any combination of exams that would result in you having to answer any more than 225 questions during any single exam session. As a worst-case scenario, this will allow you only slightly more than one minute to answer each question.

CBT Exams

Unlike the ASE paper and pencil exams, each individual ASE CBT exam has a fixed time limit. Individual exam times will vary based upon exam area and will range anywhere from a half hour to two hours. You will also be given an additional 30 minutes beyond what is allotted to complete your exams to ensure you have adequate time to perform all necessary check-in procedures, complete a brief CBT tutorial, and potentially complete a post-test survey.

Similar to the paper and pencil exams, you can register for and take multiple CBT exams during one testing appointment. The maximum time allotment for a CBT appointment is four and a half hours. If you happen to register for so many exams that you will require more time than this, your exams will be scheduled into multiple appointments. This could mean that you have testing on both the morning and afternoon of the same day, or they could be scheduled on different days, depending on your personal preference and the test center's schedule.

It is important to understand that if you arrive late for your CBT test appointment, you will not be able to make up any missed time. You will only have the scheduled amount of time remaining in your appointment to complete your exam(s).

Also, while most people finish their CBT exams within the time allowed, others might feel rushed or not be able to finish the test, due to the implied stress of a specific, individual time limit allotment. Before you register for the CBT exams, you should review the number of exam questions that will be asked along with the amount of time allotted for that exam to determine whether you feel comfortable with the designated time limitation or not.

Summary

Regardless of whether you are taking a paper and pencil or CBT exam, as an overall time management recommendation, you should monitor your progress and set a time limit you will follow with regard to how much time you will spend on each individual exam question. This should be based on the total number of questions you will be answering.

Also, it is very important to note that if for any reason you wish to leave the testing room during an exam, you must first ask permission. If you happen to finish your exam(s) early and wish to leave the testing site before your designated session appointment is completed, you are permitted to do so only during specified dismissal periods.

UNDERSTANDING HOW YOUR EXAM IS SCORED

You can gain a better perspective about the ASE certification exams if you understand how they are scored. ASE exams are scored by an independent organization having no vested interest in ASE or in the automotive industry.

Each question carries the same weight as any other question. For example, if there are 50 questions, each is worth 2 percent of the total score. Your exam results can tell you

- Where your knowledge equals or exceeds that needed for competent performance, or
- Where you might need more preparation.

Your ASE exam score report is divided into content "task" areas; it will show the number of questions in each content area and how many of your answers were correct. These numbers provide information about your performance in each area of the exam. However, because there may be a different number of questions in each content area of the exam, a high percentage of correct answers in an area with few questions may not offset a low percentage in an area with many questions.

It should be noted that one does not "fail" an ASE exam. The technician who does not pass is simply told "More Preparation Needed." Though large differences in percentages may indicate problem areas, it is important to consider how many questions were asked in each area. Since each exam evaluates all phases of the work involved in a service specialty, you should be prepared in each area. A low score in one area could keep you from passing an entire exam.

There is no such thing as average. You cannot determine your overall exam score by adding the percentages given for each task area and dividing by the number of areas. It does not work that way because there generally are not the same number of questions in each task area. A task area with 20 questions, for example, counts more toward your total score than a task area with 10 questions.

Your exam report should give you a good picture of your results and a better understanding of your strengths and areas needing improvement for each task area.

If you fail to pass the exam, you may take it again at any time it is scheduled to be administered. You are the only one who will receive your exam score. Exam scores will not be given over the telephone by ASE nor will they be released to anyone without your written permission.

SECTION 3

Types of Questions on an ASE Exam

Understanding not only what content areas will be assessed during your exam, but how you can expect exam questions to be presented, will enable you to gain the confidence you need to successfully pass an ASE certification exam. The following examples will help you recognize the types of question styles used in ASE exams, and assist you in avoiding common errors when answering them.

Most initial certification tests are made up of between 40 and 80 multiple-choice questions. The five-year recertification exams will cover the same content as the initial exam; however, the actual number of questions for each content area will be reduced by approximately one-half. Refer to Section 4 of this book for specific details regarding the number of questions to expect to receive during the initial Service Consultant (C1) certification exam.

Multiple-choice questions are an efficient way to test knowledge. To correctly answer them, you must consider each answer choice as a possibility, and then choose the answer choice that *best* addresses the question. To do this, read each word of the question carefully. Do not assume you know what the question is asking until you have finished reading the entire question.

About 10 percent of the questions on an actual ASE exam will reference an illustration. These drawings contain the information needed to correctly answer the question. The illustration should be studied carefully before attempting to answer the question. When the illustration is showing a system in detail, look over the system and try to figure out how the system works before you look at the question and the possible answers. This approach will ensure you do not answer the question based upon false assumptions or partial data, but instead have reviewed the entire scenario being presented.

MULTIPLE-CHOICE QUESTIONS

The most common type of question used on an ASE exam is direct completion, which is more commonly referred to as a multiple-choice style question. This type of question contains three "distracters" (incorrect answers) and one "key" (correct answer). When the questions are written, effort is made to make the distracters plausible to draw an inexperienced service consultant to inadvertently select one of them. This type of question gives a clear indication of the service consultant's knowledge.

Examples of this type of question would appear as follows:

1. A service consultant has just completed compiling and writing up a customer's concerns. Which of these should she do next?

   A. Confirm the accuracy of the information on the repair order.

   B. Arrange the ride home for the customer.

   C. Offer an estimate for the repairs.

   D. Have the porter wash the car.

Answer A is correct. The Service Consultant should always confirm the accuracy of the repair order prior to letting the customer leave the write-up area.

Answer B is incorrect. Arranging transportation for the customer would be done after confirming the accuracy of the repair order.

Answer C is incorrect. An estimate is not typically provided until the vehicle is diagnosed.

Answer D is incorrect. The car would not be washed until the repairs have been completed.

2. A customer calls and states that her vehicle has an electrical problem that has been recurring even after three attempts to repair it. Which of the following should the service consultant do first?

   A. Check the repair history to see if this shop has worked on it before.

   B. Offer to diagnose the vehicle free of charge.

   C. Ask the customer to explain in detail what all has been done to the vehicle.

   D. Explain that some problems require several attempts to fix.

Answer A is correct. A service consultant should determine if repairs are comeback/repeat repairs during the initial communication with the customer if possible.

Answer B is incorrect. A service consultant should never volunteer to give a diagnosis without charging for it during the initial phone conversation.

Answer C is incorrect. A service consultant should never ask for a long/detailed explanation over the phone because the customer will not be able to accurately remember each detail. It is a good idea to have her bring any service records (from other repair shops) with her when she comes to the repair shop.

Answer D is incorrect. A service consultant should never state that repeat repair attempts are normal.

SERVICE CONSULTANT A, SERVICE CONSULTANT B QUESTIONS

The question style that is most popularly associated with an ASE exam is the "Service Consultant A says ... Service Consultant B says ... Who is right?" type of question. In this type of question, you must identify the correct statement or statements. To answer this type of question correctly, you must carefully read each service consultant's statement and judge it on its own merit to determine if the statement is true.

Sometimes this type of question begins with a statement about some analysis or repair procedure. This is often referred to as the stem of the question and provides the setup or background information required to understand the conditions on which the question is based. This is followed by two statements about the cause of the concern, customer procedure, proper inspection, identification, or repair choices. You are asked whether the first statement, the second statement, both statements, or neither statement is correct. Analyzing this type of question is a little easier than the other types because there are only two ideas to consider, although there are still four choices for an answer.

Service Consultant A, Service Consultant B questions are really double true-or-false questions. The best way to analyze this type of question is to consider each service consultant's statement separately. Ask yourself, "Is A true or false? Is B true or false?" Once you have completed an individual evaluation of each statement, you will have successfully determined the correct answer choice for the question, "Who is correct?" An important point to remember is that an ASE Service Consultant A, Service Consultant B question will never have Service Consultants A and B directly disagreeing with each other. That is why you must evaluate each statement independently.

An example of a Service Consultant A/Service Consultant B style question looks like this:

1. A vehicle is in the repair shop with a complaint of poor heater performance. Service Consultant A says that the engine cooling system may need to be diagnosed. Service Consultant B says that a stuck heater control valve could be the cause. Who is correct?

   A. A only

   B. B only

   C. Both A and B

   D. Neither A nor B

Answer A is incorrect. Service Consultant B is also correct.

Answer B is incorrect. Service Consultant A is also correct.

Answer C is correct. Both Service Consultants are correct. The heater system on a vehicle works in conjunction with the engine cooling system, so it would need to be checked when there is a heater concern. Many heater systems use a heater control valve, which regulates hot coolant into the heater core when the temperature control lever is moved to the heat position.

Answer D is incorrect. Both Service Consultants are correct.

EXCEPT QUESTIONS

Another type of question used on ASE exams contains answer choices that are all correct except for one. To help easily identify this type of question, whenever they are presented in an exam, the word "EXCEPT" will always be displayed in capital letters. With this type of question, the one incorrect answer choice will actually be counted as the correct answer for that question. Be careful to read these question types slowly and thoroughly, otherwise, you may overlook what the question is actually asking and answer the question by selecting the first correct statement.

An example of this type of question would appear as follows:

1. All of the following information should be on the customer appointment log EXCEPT:

   A. Customer name

   B. Estimated time of repair

   C. Vehicle color

   D. Vehicle year, make, and model

Answer A is incorrect. The customer name should always be on the appointment log.

Answer B is incorrect. A rough time of repair estimate should be on the customer appointment log. This time will usually not be extremely accurate, but it helps to plan the number of vehicles that need to be scheduled for each day.

Answer C is correct. The vehicle color is not a typical piece of information that should be on the customer appointment log.

Answer D is incorrect. The vehicle year, make and model should usually be present on the customer appointment log.

LEAST LIKELY QUESTIONS

LEAST LIKELY questions are similar to EXCEPT questions. For this type of question, look for the answer choice that would be the LEAST LIKELY cause (most incorrect) for the described situation. To help easily identify this type of question, whenever they are presented in an exam the words "LEAST LIKELY" will always be displayed in capital letters. Read the entire question carefully before choosing your answer.

An example of this type of question would appear as follows:

1. Which of the following repair procedures would be the LEAST LIKELY to be considered a high priority repair?

   A. Replacement of a worn tie rod end

   B. Replacement of the rear wiper blade

   C. Replacement of a tire with the steel showing

   D. Replacement of front brake pads

Answer A is incorrect. A worn tie rod end is a high priority repair since it could come loose and cause the driver to lose control of the vehicle.

Answer B is correct. A worn rear wiper blade would not cause the vehicle to be dangerous to drive.

Answer C is incorrect. A severely worn tire would be a great danger to continue using. This repair would need to be made immediately.

Answer D is incorrect. Worn brake pads would make the vehicle dangerous to drive so they would need to be replaced very soon.

SUMMARY

The question styles outlined above are the only ones you will encounter on any ASE certification exam. ASE does not use any other types of question styles, such as fill-in-the-blank, true/false, word-matching, or essay. ASE also will not require you to draw diagrams or sketches to support any of your answer selections, although any of the above described questions styles may include illustrations, charts, or schematics to clarify a question. If a formula or chart is required to answer a question, it will be provided for you.

SECTION 4

Task List Overview

INTRODUCTION

This section of the book outlines the content areas or *task list* for this specific certification exam, along with a written overview of the content covered in the exam.

The task list describes the actual knowledge and skills necessary for a service consultant to successfully perform the work associated with each skill area. This task list is the fundamental guideline you should use to understand what areas you can expect to be tested on, as well as how each individual area is weighted to include the approximate number of questions you can expect to be given for that area during the National Institute for Automotive Service Excellence (ASE) certification exam. It is important to note that the number of exam questions for a particular area is to be used as a guideline only. ASE advises that the questions on the exam may not equal the number specifically listed on the task list. The task lists are specifically designed to tell you what ASE expects you to know how to do and to help prepare you to be tested.

Similar to the role this task list will play in regard to the actual ASE exam, Delmar, Cengage Learning has developed six preparation exams, located in Section 5 of this book, using this task list as a guide. It is important to note that although both ASE and Delmar, Cengage Learning use the same task list as a guideline for creating these test questions, none of the test questions you will see in this book will be found in the actual, live ASE exams. This is true for any test preparatory material you use. Real exam questions are *only* visible during the actual ASE exams.

Task List at a Glance

The Service Consultant (C1) task list focuses on three core areas, and you can expect to be asked approximately 50 questions on your certification exam, broken out as outlined:

A. Communications (26 questions)

   1. Customer Relations (12)
   2. Sales Skills (10)
   3. Internal Relations (4)

B. Product Knowledge (21 questions)

   1. Engine Systems (Includes Mechanical, Cooling, Fuel, Ignition, Exhaust, Emissions Control, and Starting/Charging) (4)
   2. Drive Train Systems (Includes Manual Transmission/Transaxles, Automatic Transmission/Transaxles, and Drive Train Components) (3)
   3. Chassis Systems (Includes Frame, Brakes, Suspension, Steering, Wheels, Tires, and TPMS) (4)
   4. Body Systems (Includes Body Components, Glass, Heating and Air Conditioning, Electrical, Restraint, and Accessories) (3)

5. Services/Maintenance Intervals (3)
6. Warranty, Service Contracts, Service Bulletins, and Campaign/Recalls (2)
7. Vehicle Identification (2)

C. Shop Operations (3 questions)

Based upon this information, the graph shown here is a general guideline demonstrating which areas will have the most focus on the actual certification exam. This data may help you prioritize your time when preparing for the exam.

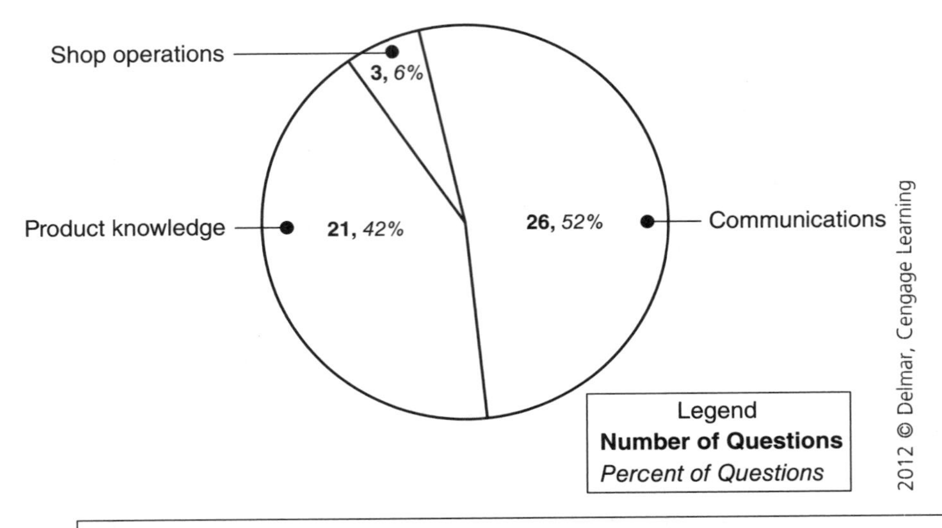

> *Note:* The actual number of questions you will be given on the ASE certification exam may vary slightly from the information provided in the task list, as exams may contain questions that are included for statistical research purposes only. Do not forget that your answers to these research questions will not affect your score.

SERVICE CONSULTANT (TEST C1) TASK LIST

A. Communications (26 Questions)

The ability to easily communicate is without a doubt the most important skill that a service consultant needs to possess. Written and oral communications are both very important. The sections that follow will assist the service consultant in understanding the various levels and types of communication that must take place in an automotive repair facility.

1. Customer Relations (12 Questions)

1. Demonstrate proper telephone skills.

Using the telephone properly includes greeting customers, speaking clearly, and demonstrating courtesy when taking calls or messages. Because of the varying environments in which service consultants work, questions addressing this “soft skill” are based on common sense. Since the first word of the job title is Service, use it as a guide on all areas of this test, excluding the product knowledge section.

2. Obtain, confirm, and document pertinent vehicle/ customer contact information.

One of the most important communication skills the service consultant must perform is collecting the information needed to create a complete work order. Correct vehicle information is paramount throughout the repair process. Many updates and recalls are VIN-specific, getting the correct parts is vehicle-specific, and even making sure that the right vehicle gets the right repair is driven by the repair order. Vehicle information includes year, make, model, powertrain information, VIN, production date, body configuration, and emissions or options information. Even items such as whether it is a 2-door or 4-door, or has power windows, or even what color a car is, can become essential.

3. Identify, verify, and document customer concern/request.

Interviewing the customer to determine their concerns is a critical skill in effective service consulting. Properly documenting the conditions and descriptions of each concern should give the technician a good starting point for making a diagnosis. Simply writing down what the customer says is not enough. The skilled service consultant must be able to help the customer remember seemingly unrelated facts that may make the difference in whether a diagnosis is successful or unsuccessful. This is done by asking questions that encourage the customer to verbalize the symptoms of their concern. Asking questions that start with "when," "how often," "where," and "to whom" can really help to fill in the question of "why." The service consultant should first address the root cause for which the customer made the visit. This task also encompasses the collection of requested general service needs, like oil change or tire rotation. Questions throughout the customer relations area frequently overlap tasks in other sections. This is not a mistake; it is ASE's effort to test in different, real-world scenarios.

4. Demonstrate appropriate greeting skills.

Greeting skills are the things we do and say to demonstrate to our customers that they are welcome. Shaking hands, smiling, verbal and non-verbal communications are all examples of greeting skills. ASE tests are built on industry standard procedures. Smiling and offering a handshake are common courtesy, common sense, and industry standard. Another similar example of a greeting skill would include asking the customer for their name. Eye contact is most important. People go for service from people they like and trust. Money can become the last issue.

5. Discuss alternative transportation options.

One of the things service consultants are called on to do daily is resolve transportation needs for their customers. Options may include car rental or, at minimum, a ride to home or work. The focus of this task is to test the understanding of the customer service and business liability side of the equation. The shop might not have loaner vehicles, but it should be apparent that allowing an uninsured or unlicensed customer to drive a shop-owned vehicle would be a problem. Here, as in most of the soft skills areas, the ability to apply common sense is the most invaluable tool the service consultant has.

6. Promote procedures, benefits, and capabilities of service facility.

A key to customer retention is their understanding and perception of *value* in the work performed for them. A good service consultant sets his facility apart from the crowd by explaining the benefits of services and the skill level of the shop's technicians. The service consultant must understand the difference between features and benefits as applied to this test.

An example of a feature might be longer-lasting spark plugs. The benefit is that the customer will not have to replace them as often. Most savvy customers buy benefits. An example of a feature with no real benefit is an extended warranty sold on a new car for 60,000 or 75,000 miles. This sounds like an added benefit, but the truth is that much is covered already by the standard warranty. Very little is extended. This is an example of selling on features. At one point, such a marketing ploy might have helped sell some cars, but most customers now look for a feature to provide them with a convenience or some perceived value. Think of a benefit as something that provides a value in terms of time, money, or convenience.

7. Check vehicle service history.

Vehicle history can mean the difference between a free repair and one a customer pays for. It can be an invaluable tool to help technicians in the diagnostic process. Questions under this task will be very general due to the myriad ways that service facilities go about gathering history. Ignoring history is a big mistake for a service consultant. Computerized systems do a great job of tracking histories.

8. Identify and recommend service and maintenance needs.

There are many possible ways to identify service and maintenance needs. The service consultant may not use or have access to all of them, but he/she should be aware of them. The most basic method is looking at the vehicle's odometer. Customer history can play in here. The vehicle's owner manual contains much of the maintenance information. Manufacturer and aftermarket information systems and websites, as well as the National Highway Traffic Safety Administration (NHTSA), are some electronic reference resources. The last and probably most viable source is the technician who conveys maintenance needs on the repair order.

9. Establish job status/completion expectations.

In a recent American Automobile Association (AAA) survey, the number one complaint customers had with repair shops was that their vehicle's service was not completed when it was promised. Anticipate that questions will address insuring and communicating completion expectations. For many service consultants, promising a completion time at the time of drop off is standard operating procedure. The industry is sending the message that the best communication occurs after the technician evaluates the vehicle. He or she must update the customer throughout the repair and avoid overpromising.

10. Confirm the accuracy of the repair order and obtain repair authorization.

Taking into consideration estimate laws in various states, this area of the test revolves around providing the customer with pricing, asking for and closing the sale, and documenting authorization for needed repairs. To enter or drive the car in the shop is a risk without the customer's signature.

After gathering all of the necessary data from the customer, the service consultant should verify the accuracy of the data with the customer. This verification can be done by reading the pertinent information aloud back to the customer or by letting the customer read the screen before printing the repair order.

11. Identify customer types (first-time, warranty, repeat repair, fleet, etc.) and method of payment.

The service consultant must ask questions to determine the type of customer they are dealing with. Typically, this will be determined when making the appointment. Good questions should be asked when scheduling the appointment to help plan how to handle the customer. Some of the various categories of customers include first-time, warranty, repeat repair, and fleet. Understanding the category with which each customer aligns will assist the service consultant and the repair shop in providing the best experience possible for each customer.

The payment policy for the repair shop should be in plain sight of the trouble ticket write-up area. If the customer is visiting the shop for the first time, it is wise to explain the acceptable payment methods during the write-up process.

12. Present professional image.

Presenting a professional image can vary from how a person dresses to the way a person treats or talks to customers and other employees. The service consultant is often the only example the customer has to generate an opinion of the quality of the facility. As the point person for the business, the likelihood of a customer returning to the facility by choice rests more on the service consultant's shoulders than anyone else's. The grooming of the facility is also a key component of presenting a professional image to the customer. All of the employees of a repair shop should strive to present the best image possible by wearing appropriate attire that is clean and neat. In addition, the staff should strive to keep the facility as clean as possible.

13. Perform customer follow-up.

Customer follow-up is a critical element of customer service communication and is an area to which the service consultant should give his/her attention. It can range from making a call to a customer to thank him for bringing his vehicle in to taking the responsibility of calling when a special-order part arrives.

Some repair shops make it a practice to perform a follow-up call after every repair. This service allows customers to rate several areas of their service experience. Repair shops can use this data to see feedback on what customers like about the repair shop as well as what customers do not like. A progressive business is always looking for ways to improve the service that they provide.

14. Explain and confirm understanding of work performed, charges, and warranties.

The service consultant is responsible for controlling all customer contact so that a complete understanding is reached. The time saved with a quick pick-up can be totally offset by the angry follow-up that occurs when the customer's expectations do not match the actual work performed. The message here is that five minutes spent with customers early on can save hours of headache, or even prevent the loss of a customer from the shop. The pick-up is the last—and sometimes the only—opportunity the service consultant has for a face-to-face marketing effort. When possible try to schedule pick-up times. This can avoid the stress that develops when all of the customers show up at once.

A professional service consultant is also aware of any pertinent warranties that each repair procedure carries with it. For example, if a battery was replaced, then the service consultant would need to explain the warranty period on the new battery.

2. Sales Skills (10 questions)

1. Provide and explain estimates.

After the service consultant has assembled a complete and accurate estimate for work needed on a customer's vehicle, it is time to present it to the customer by phone or in person. The best time to reach a full understanding with a customer is before the work is done. It is an excellent approach to ask the customer how much information she would like provided about the work estimated. Some customers may want considerable detail and explanation; other customers may only want summary information. Be willing to adapt the presentation to the customer's personality type and knowledge level.

Many service consultants provide incorrect information because they are in a hurry. This will almost always start a web of confusion that will end in the customer's distrust of the service consultant and, ultimately, the repair facility. It is of critical importance that the items on the repair order include all costs associated with them. If the service consultant offers an estimate that does not include shop supplies, environmental charges, tax, etc., the customer will perceive this as dishonest. Selling the estimate to the customer is as much an effort to sell yourself and the shop's professionalism as it is selling the work. People buy service from people, not from businesses. When possible, try to stagger work drop-off and delivery times. This helps to ensure that there will be enough time to explain all that is necessary.

2. Identify and prioritize vehicle needs.

When preparing an estimate, it is important to keep in mind which services are absolutely critical now, which are discretionary, and which are more cost-effective when grouped with other services. Many customers do not have the discretionary funds to perform all of the services recommended at one time. The service consultant has the opportunity to become the hero by helping to point out savings with bundled services or items that can be rescheduled for a later date. Sometimes the business has extended financing options that can make it possible to do all of the work at one time. Remember, control the conversation, but give the customer options. The service consultant does this every day and develops experience with financial and repair strategies.

3. Address original concerns with customer.

When a service consultant offers complete explanations and solutions to the customer's concerns on the repair order, he/she is addressing customer concerns. Another example of addressing customer concerns is when customers have questions about the reason and expectations from a given repair. Addressing a customer's concerns is never a negative; it is an opportunity to make that customer feel at ease with the service consultant, the business, and the work to be performed. A professional service consultant should treat customers with honesty, fairness and respect.

4. Communicate the value of related and additional services.

The service consultant must help the customer understand the benefits of services that will really complete a job. Most people dislike being without their car, which leads them to bring it in for servicing only when absolutely necessary. It is important for the repair business to keep this in mind and offer and anticipate needed services while a vehicle is in for other repairs. This will help keep the customer on the road, which is a win for both the customer and the shop. It also helps avoid failed parts or additional repairs due to incompleteness of the original sale and repair.

Just as the service consultant needs to learn how the many vehicle systems interrelate to one another, he/she must also help the customer understand why it is advantageous to group some services together in order to save time and money. Reliable service should be the selling point for keeping the maintenance items up to date.

5. Explain product/service features and benefits.

Features are the list of items that define a product, while *benefits* are the tangible or perceived items the customer takes along with the product. In the communications section, the difference between features and benefits was explained. A feature of new spark plugs might be that they have a long-lasting design. The benefit is that the customer will not have to replace them as often. The plug's extended life is of no consequence unless there is a perceived or tangible value to that long life. That value comes from the fact that the expense of replacing spark plugs, as well as the inconvenience of more frequent visits to the shop, is reduced because of the design feature. Therein lies the benefit. When a service consultant sells products or services to customers, be sure that the benefits are included in the sales pitch. Another way to think of it is that the benefit is what the product does for the customer or how it makes the customer feel.

6. Overcome objections/finalize sale.

There may be several reasons for a customer's objection to a needed repair. It is up to the service consultant to discover those reasons. Usually the best approach is to ask the customer why she does not want the service performed. It is possible that she has had it done just recently, which is something the service consultant should know. Nothing undermines a service consultant's credibility more than recommending a repair that has just been done (or that the customer thinks has just been done).

Another common reason for an objection is that the customer does not understand the operation, or he fails to see the value in it. Asking questions to find out what the concern is will help them understand. The service consultant still may not make the sale, but the fact that he/she has made the effort to explain the position goes a long way toward building customer trust. History tracking or asking some initial qualifying questions can help avoid service suggestion pitfalls.

If the customer suggests he would like a second opinion, a good response might be, "Yes, I think it is always a good idea to get a second opinion. I think our diagnosis and pricing are fair and consistent with other repair facilities, but I would be happy to offer you a printout of our diagnosis to have another shop confirm it for you." Most customers will decide that if the service consultant is confident enough to allow another shop to confirm this diagnosis, they might as well have the work done now as opposed to going through the cost and time to do another diagnosis. Trust is key to the relationship between the consultant and the customer.

3. *Internal Relations (4 Questions)*

1. Effectively communicate customer service concern/request.

Collecting all the information in the world from Task A.1.3 is of no value at all unless the service consultant can make sure that the information collected is useful to the technician and is clearly communicated. "Effectively" can also mean communicating without having to call the customer back several times. Expect questions in this task to focus on the communication between the service consultant and the technician relating to work orders. In the future, it may be interpreted to include verbal communication. Do not add or delete any information.

2. Understand the technician's diagnosis and service recommendations.

Arguably, the real craft of service consulting is the ability to take the diagnosis and recommendations of a technician and turn it into a coherent description of the cause and correction needed to satisfy and alleviate the customer's concern. This area is really about the service consultant's product knowledge applied to practical work. For most service consultants, this means asking questions of technicians to gain an understanding of how the various systems work. The most successful service consultants do an excellent job of taking the technical information and offering an accurate but customer-friendly explanation of the problem. Remember, most customers will not want an in-depth technical explanation. Keep it simple and focus on the issue that concerns the customer.

3. Verify availability of required repair parts.

To avoid problems in the area of parts availability, the skilled service consultant confirms the availability and timing of parts before promising a return time to the customer or directing the technician to begin work. This task has a sort of procedural aspect to it, so expect to use good common sense here.

Do not expect to see the use of purchase orders or any specific parts system in here since not all businesses are the same. Do expect that scenarios that might occur in relation to a parts department will be posed as questions on the test. Communication is key in this area.

4. Establish completion expectations.

Service consultants who work with the technicians are more able to determine the completion time of any given task. Many consultants have a very good knowledge and working relationship with their technicians and can plan workflow with excellent results. Most technicians would prefer to be given vehicles for servicing in the order that they need to be returned to their owners. All of the other factors, such as customer need, parts availability, and technician/equipment availability play into this. Some jobs require different skills, which should also be considered for job distribution. The consultant must take the lead, whether working with shop supervisors or technicians, to offer customers realistic completion times.

5. Monitor repair progress/quality control.

During the course of the day, maintaining a hands-on approach to workflow keeps the shop's productivity high. The service consultant will have more options if problems that come up can be addressed and communicated to the customer as early as possible. The service consultant may find that a walk through the shop to gather updates from the technicians can be a simple and effective way to monitor repair progress. Some electronic dispatching systems allow members of the team to update their progress throughout the day.

The service consultant needs to keep an eye on the quality of the repairs that are being completed in the shop. When a customer has to bring his vehicle back to the shop for the same problem, the term is called a *comeback*. The service consultant should closely investigate each comeback case and see what caused this situation to happen. Some of the causes for comeback cases include technician error, poor quality parts, and miscommunication. No matter the cause, the service consultant needs to try to make decisions and recommendations that will reduce the number of comeback cases.

6. Document information about services performed or recommended.

When the shop performs a service, it is critical to document information in such a way that the technician can review notes and know what work was performed and why. It is also important to document in such a way that the customer understands and can perceive value in the service provided. This may include maintaining internal documents along with final customer invoices, offering references to the Transportation Safety Board (TSB), or providing marketing campaign information relevant to the work performed. If a customer declines to have a recommended service performed, the service advisor should have the customer sign a document stating that she chose to not follow the recommendation. This practice will protect the shop in case the vehicle fails in an area related to the recommendation.

7. Communicate with shop personnel about shop production/efficiency.

The difference between productivity and efficiency, as well as how to keep things humming along, is the focus of this task. *Productivity* is the amount of work a shop gets out the door in a given period of time, while *efficiency* is a measurement of how effectively work is completed. If a technician is at work for eight hours, receives eight hours of sold labor, and completes it in eight hours, he is 100 percent efficient and 100 percent productive. If, on the other hand, he is at work for eight hours, has six hours of sold labor, and completes the work in four hours, he is 150 percent effective, but only 75 percent productive. This information provides an invaluable tool for tracking technicians.

In the first scenario, the technician and the service consultant are working in harmony. The only way to make more in this scenario is for the technician to have higher efficiency so that more work can be completed in the same amount of time. This often calls for scheduling some highly efficient work that can be performed faster than the "book" time because of the technician's experience and skill. This relation between productivity and efficiency is critical to the profit of the shop and financial success of the employee.

In the second scenario, the technician is working efficiently, but does not have enough work to be as productive, in terms of service fulfilled, as he or she could be. If the numbers are shuffled around, it is easy to see that this technician could produce 12 hours a day of sold or billed labor if the work were there. Questions about this task will be pretty general, as most service consultants in the original equipment (OE) dealer networks are not called on to perform this kind of analysis. This is invaluable information to understand in any service environment, however, because it can make the difference between happy technicians/profitable shops and the revolving employee/business failure scenarios.

8. Maintain open lines of communication within the organization.

The last task in the internal relations section may well be the most ambiguous, since maintaining open lines of communications could encompass every and any kind of communication in the business. Most of the questions written to this task will involve interpersonal communication between individuals. This might include verbal, written (memo), or telephone messages. Expect that problem resolution will be a popular subject. Small meetings daily, weekly, or monthly can help in maintaining open lines of communication so every employee has a chance to contribute to finding remedies to problems and sharing successes.

B. Product Knowledge (21 Questions)

Sections B.1 through B.4 concern technical knowledge of vehicle systems. Since each section has the same three tasks, we will address all three tasks at the same time. The tasks are:

1. Identify major components and location.
2. Identify component function.
3. Identify related systems.

Each section is organized with charts and diagrams to help provide a working knowledge of each system, including how they relate to other systems. The first task is to identify the location of major components. Vehicles have so many design variations it would be nearly impossible to write a question that would always fit every vehicle built.

This guide will use the word "usually" in the product knowledge section for this reason. ASE test questions do not use words like "always" because of this rule. The service consultant is not expected to possess the depth of technical knowledge that a technician must, but a working knowledge of system functions and interrelationships is necessary to provide good customer service and to communicate with technicians and parts people. The second and third tasks of identifying the component function, as well as identifying related systems, are covered in the charts and diagrams for each section. The information about each system is covered on a basic level in order to give a service consultant a basic understanding of each area of the vehicle.

1. Engine Systems (4 Questions) (Includes mechanical, cooling, fuel, ignition, exhaust, emissions control, and starting/charging)

An integrated approach to each system indicated on the task list should help assist service consultants gain a better understanding of the components, their functions, and their relationships to other systems. For each number in the exploded diagrams, there is a corresponding numbered section in the chart following it. Understanding this information will improve the technical knowledge base needed to answer any questions on the Service Consultant Test.

Mechanical

The valve train is the system that mechanically opens and closes the valves in the engine. There are several different methods of operating the valve train.

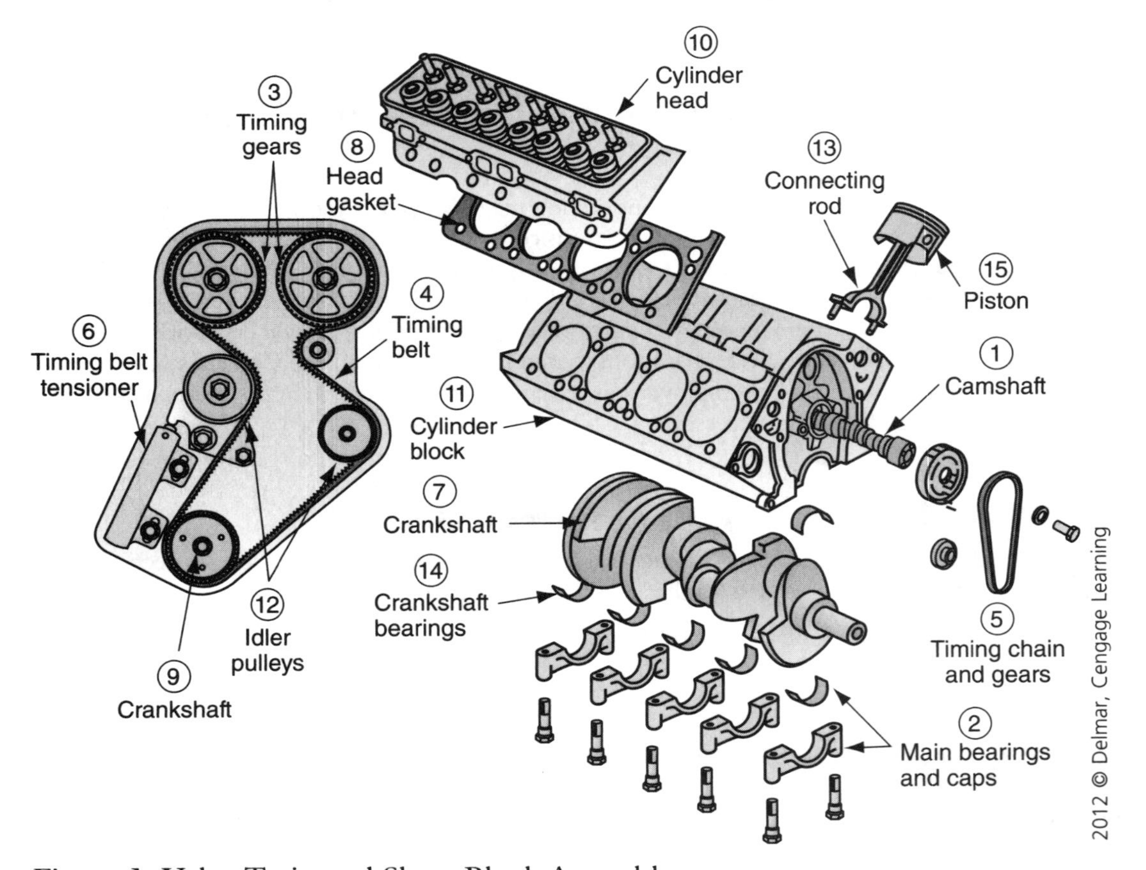

Figure 1. Valve Train and Short Block Assembly

Valve Train		
Diagram Number	**Component Name**	**What It Does**
1	Camshaft	The *camshaft* has eccentric lobes that rise from the centerline of the shaft to push on valve lifters or rocker arms and cause valves to open. The cam is driven at half crank speed by the timing chain, timing belt, or timing gears connected to the crankshaft.
2	Main Bearings and Caps	Main bearings and caps attach the crankshaft to the block; they must be torqued to specification. A sequence of tightening may be required.
Not Pictured	Valves	*Intake valves* control the introduction of fuel and air into the engine. They usually open at the beginning of the intake stroke. *Exhaust valves* control the elimination of by-products of the combustion process. Both valves are closed during the compression and power strokes.
3	Timing Gears	*Timing gears* are driven by the timing belt or a chain. The example shown is a belt-driven design dual-overhead camshaft. This gear is often referred to as a *sprocket*.
4	Timing Belt	*Timing belts* are made of rubber and have reinforcing bands and teeth, also made of rubber, to mesh with cam and crank gears. Many engines have a valve train, which is known as *interference*. This means that if the timing belt breaks, the valves will come in contact with the pistons. This is why it is critical to follow the mileage intervals recommended for belt replacement. It is still important in non-interference engines (engines that will not have valve-to-piston contact if the belt breaks) to follow the intervals to avoid a breakdown and resulting tow to the shop.
5	Timing Chain and Gears	*Timing chain and gears* perform the same function as a timing belt. They are not a maintenance item like the timing belt, however; they are internal components that are lubricated by the engine's oiling system.
6	Timing Belt Tensioner	*Tensioners* provide proper pressure on the timing belt or, in some cases, timing chain. They may be spring-loaded, manually adjusted, or actuated by engine oil pressure.
Not Pictured	Camshaft Pulley Seals	Because the camshaft must have oil to lubricate it when the cam drive is a belt drive, there are

(Continued)

Valve Train		
Diagram Number	**Component Name**	**What It Does**
		seals that keep the oil in the engine and off the belt. This is a very commonly replaced component during timing belt service.
7 and 9	Crankshaft	The *crankshaft* is the rotating member that provides power output. All pistons are connected by a connecting rod and bearing. The crankshaft is a large offset shaft. The journals are configured in 60-, 90-, or 180-degree relationships depending on the type of engine. The large weights hanging opposite the journals are used to counterbalance the weight of the journal and the rod and piston assembly that attaches to each journal. In most V6 or V8 engines, there are two connecting rods bolted Siamese to each crank journal, while 4-cylinder engines usually have one rod on each crank journal. This is a heavy component made of steel or nodular iron in most cases. This rotates at twice the speed of the camshaft.
8	Head Gasket	The *head gasket* provides a positive seal between the head and cylinder block. Oil, antifreeze, and combustion are all sealed by this component part.
Not Pictured	Front Crank Seal	This component functions the same as a cam seal. A front crank seal is used in all engines regardless of valve train configuration. This seal often fails and is replaced during timing belt service.
Not Pictured	Valve Springs and Retainers	These components apply pressure to keep the opening and closing of the valve in matching motion to the lobes of the camshaft and to hold them closed.
Not Pictured	Timing Cover	The *timing cover* protects the timing belt. In timing chain applications, it covers the timing chain, controls oil, and often contains the oil pump.
Not Pictured	Valve Cover	The *valve cover* attaches to the cylinder head and covers the valve train components.
10	Cylinder Head	The *cylinder head* holds all of the valve train components and has air flow ports for the exhaust and intake. It bolts onto the cylinder block, with the intake and exhaust manifolds bolted to it. The complete package is often called the *induction system*.

Short Block Assembly		
Diagram Number	**Component Name**	**What It Does**
11	Cylinder Block	The cylinder block is the backbone of the engine: All major components bolt to it. The cylinder bores are the large holes in it. The pistons go in these holes. The crankshaft rotates in bearing inserts in the block's main saddles.
12	Idler Pulleys	These components provide direction change in the belt configuration. They are in a fixed mounting and simply "idle" in rotation when the belt is in rotation.
13	Connecting Rod	The *connecting rod has* two holes in it when viewed by itself. The large hole is the side that holds the rod bearing and splits apart to bolt to the crank journal. The small end accepts the wrist pin of the piston, which is the point the piston pivots on when the crankshaft is spinning.
14	Crankshaft Bearings	The *bearings* in modern engines are a composite of different materials clad together. The back shell is usually aluminum with a copper layer on it, and then a soft metal alloy, similar to lead, that is the actual bearing surface. The bearing does not have any moving parts like other bearings. In the engine, its job is to carry an oil film to the surfaces of the crank and maintain adequate clearance between moving parts. When the clearances are excessive, knocking noise and loss of oil pressure can result. If the bearing clearance is too tight, the oil film that cools and protects the parts is too thin to be effective, which will cause damage to the parts.
Not Pictured	Oil Pump	The *oil pump* is the heart of the engine. Oil pumps may be driven directly by the crankshaft or by gears and a driveshaft from the valve train. The pump pulls oil from the oil pan and passes it to the oil filter. Because it can move more volume than the internal clearances of the engine can flow, pressure is created in the oiling system. If these clearances increase due to wear in the bearings due to age, oil flow may increase, but the pressure will be lowered. This pressure occurs in small passages inside the engine, sort of like the plumbing inside a house, carrying oil to all of the bearing surfaces in the engine. The oil pan and valve covers are

(Continued)

Short Block Assembly		
Diagram Number	**Component Name**	**What It Does**
		not under pressure. They simply keep the oil in the engine. The oil pan is a reservoir to which the oil returns. Engines use drains like the gutters on a house to return unpressurized oil back to the pan at the bottom of the engine.
Not Pictured	Rear Main Seal	This is the large seal that keeps the oil from the rear crank main bearing inside the engine. This seal may be a two-piece design that is serviced by removing the rear main bearing cap, or it may be a one-piece, full-circle seal that is installed from the outside rear of the engine, requiring transmission and/or transaxle removal.
15	Piston	The *piston* is made of aluminum in all modern engines. When it moves to the bottom of the cylinder, it creates a vacuum that pulls air and fuel into the cylinder; this is called the *intake stroke*. It then rises to the top to compress the air it just took in, called the *compression stroke*, before the spark plug sparks and causes the high-pressure fuel and air mixture to create lots of heat as it oxidizes (burns). This heat causes expansion, which forces the piston back down; this is the *power stroke*. This causes the crank to spin because of the eccentric journals discussed earlier. The crank continues to spin due to inertia. The final stroke in the 4-stroke cycle is known as the exhaust stroke. This is when the leftover by-products of combustion are expelled. Since different cycles are happening in the other cylinders, the crank is being pushed by a compression stroke from another piston continuously. Due to the four strokes needed, the crankshaft will rotate twice for every camshaft rotation. This is why the crank gear will always be the smaller, or half size, when compared to the camshaft gear.
Not Pictured	Piston Rings	There are two types of *piston rings* on each piston. One is the compression ring. There are usually two of them stacked right on top of each other in grooves in the piston. Since the piston must be lubricated to slide up and down the cylinder, it cannot be too tight in the bore. The compression rings are responsible for providing an extremely effective and small

Short Block Assembly		
Diagram Number	**Component Name**	**What It Does**
		sealing surface that requires very little lubrication. The second ring type is the oil ring, which helps to carry oil up the cylinder bore for lubrication. This ring then scrapes the oil back off and passes it through notches cut in the piston skirt on the way back down.

Cooling System

The cooling system serves two different purposes. First, it keeps the engine at a steady temperature. Second, it provides the hot water that makes the heater work. The cooling system is made up of many components. The diagram below will show a basic system common to most vehicles. The cooling system contains a mixture of antifreeze and water. The mixture carries heat from the engine components to the radiator and the heater core where the heat is exchanged by convection between the outside air and the radiator or heater core. The components of a typical cooling system are listed below.

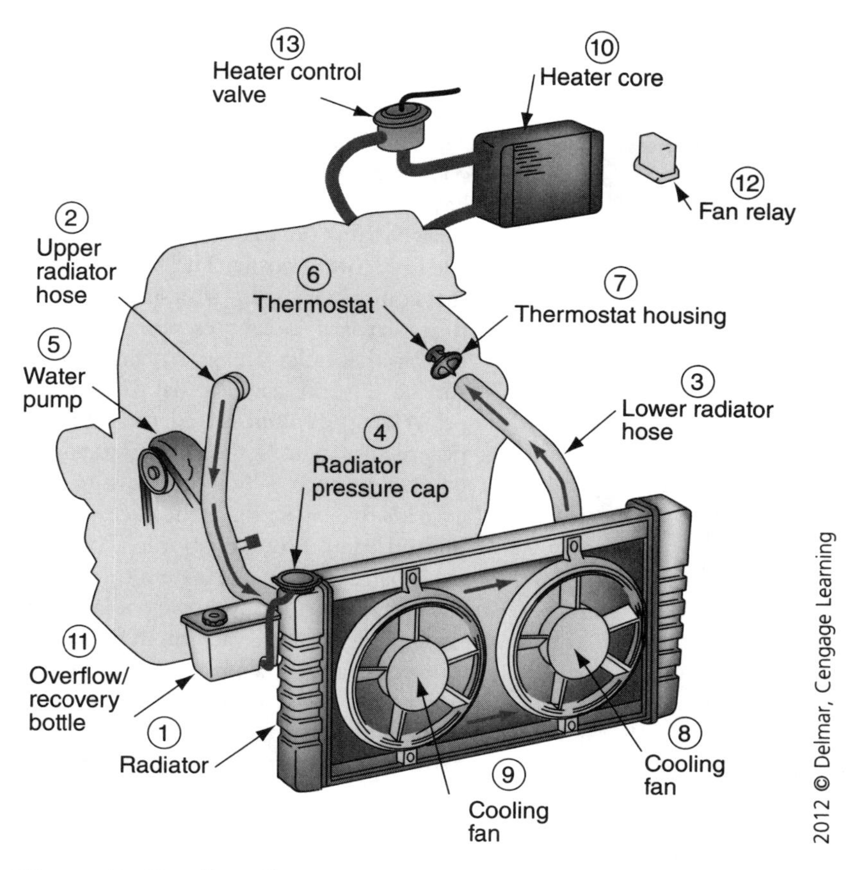

Figure 2. Cooling System

Cooling System Exploded View		
Diagram Number	**Component Name**	**What It Does**
1	Radiator	The *radiator* is a heat exchanger that is mounted at the front of the vehicle. Warm coolant enters it, the outside air moving across it removes heat from the coolant, and it returns to the engine to start the cycle over again. Radiators are carefully chosen to have the correct amount of heat transfer efficiency called British thermal units (BTUs). If the radiator is too small, becomes dirty, or is restricted, the engine will overheat due to inadequate heat exchange. It should be noted that most vehicles equipped with automatic transmissions circulate their transmission fluid through a separate heat exchanger inside the radiator to help warm the transmission up on cold days, and to maintain consistent temperature after the vehicle has warmed up.
2	Upper Radiator Hose	The *radiator hoses* are rubber hoses that provide a flexible connection between the engine and radiator.
3	Lower Radiator Hose	The radiator hoses are rubber hoses that provide a flexible connection between the engine and radiator.
4	Radiator Cap	The *radiator cap* is responsible for maintaining the cooling system's pressure. Late-model vehicles have closed cooling systems, which means that they use a radiator cap to control the coolant level along with a recovery bottle. The radiator cap has a seal that comes in contact with a surface inside the top of the radiator or coolant reservoir. There is a spring in the cap that holds pressure against this sealing surface. The pressure is usually between 13 and 16 psi. When the engine is cold, the cooling system has no pressure in it. As the engine warms up, the coolant expands. When the pressure in the system exceeds the cap's rating, the excess coolant will be pushed out to a recovery bottle. This expansion will also push any air in the system out. In turn, when the engine is shut down, the coolant retracts as it cools. This creates a vacuum in the cooling system. The coolant is drawn back out of the recovery bottle until the system is full. There are really no moving parts involved, just a controlled use of expansion and contraction. If the system is otherwise leak-free, it will be full at all times and free of air, which causes accelerated deposits that can restrict the tubes of the radiator. Another advantage

Cooling System Exploded View		
Diagram Number	**Component Name**	**What It Does**
		of running a pressurized cooling system is that coolant under pressure has a higher boiling point.
5	Water Pump	One of the radiator hoses is connected to the inlet side of the *water pump*. The pump pulls coolant from the cooled side of the radiator and pushes it out through passages that travel through internal passages in the cylinder head and block. These passages are cast around all of the hot parts of the combustion process, like the cylinders and the combustion chambers in the heads. The shaft that connects to the "paddle wheel," called an *impeller*, goes to the outside of the engine and has a pulley attached to it that is driven by a belt. Seals and bearings to allow it to spin at high speed while keeping the coolant in the engine. This is the location where most water pumps fail. Water pumps come in many shapes and sizes, but all perform the same task.
6	Thermostat	The *thermostat* for an engine is located inside the engine and is submerged in coolant. It controls the amount of coolant that enters the radiator from the engine. When the engine is cold, a big spring inside the thermostat holds it closed to keep all of the coolant flowing in the engine. As the engine reaches operating temperature, usually about 195°F (91°C), the spring in the thermostat relaxes, allowing coolant to flow to the radiator for heat exchange. If the temperature drops, the thermostat closes the opening down. This keeps the temperature consistent.
7	Thermostat Housing	Most engines have a *thermostat housing* that is either part of the engine or a separate part. They are often made of light-duty aluminum and have a radiator hose connection. They are subject to leaks and warpage in many applications.
8 and 9	Cooling Fan	Since vehicles operate under a variety of conditions there is no way to guarantee adequate air flow across the radiator, for instance, when sitting in traffic or moving at slow speeds. The *cooling fan* is used to provide the wind when there is none. Cooling fans may be old static fans that move at the same speed as the water pump they are bolted to. They may be electric and controlled by a temperature switch or engine computer, or they may be mounted on a fan clutch. The cooling fans pull/push air through the radiator. Vehicles with air conditioning use

(Continued)

Cooling System Exploded View

Diagram Number	Component Name	What It Does
		these fans to pull/push air across the condenser. The condenser is located in front of the radiator.
Not Pictured	Fan Clutch	*Fan clutches* are used in applications where the fan is mounted to a moving part of the engine, usually the water pump. Fan clutches can be thermostatic or speed sensitive. In either case, they vary the amount of air the fan moves by slowing down the fan blade relative to engine speed. This allows the use of a large, highly effective blade for slow speeds that can be slowed down to improve engine performance when it is not needed.
10	Heater Core	The *heater core* is a small heat exchanger. Coolant from the engine is circulated through it, and a blower blows air across the core to provide hot air inside the vehicle.
11	Overflow/Recovery Bottle	*Coolant recovery/overflow bottles* are the reservoirs that coolant moves into and out of as it expands and contracts. In many late-model vehicles, the bottle has been replaced by a totally closed system wherein the bottle functions as the radiator pressure and off-gassing tank to help trapped air get out of the coolant.
Not Pictured	Antifreeze	The coolant in the engine comes in many types. It can be green, red, yellow, orange, or pink. The name "*antifreeze*" only describes half its function: In addition to dropping the freezing point of water to around –40°F (–40°C), it also helps to prevent boiling up to 240°F (116°C) or more, depending on system pressure. The coolant is mixed with water in a 50/50 mixture in almost all applications. The coolant has the additional functions of lubricating the water pump (protecting the metal components of the engine) and improving the temperature transfer capabilities of the components with which it comes in contact.
12	Fan Relay	This component provides a low-current control for a high-current demand (fan motors). On modern autos, this cycle is computer-controlled, receiving inputs from many sensors.
13	Heater Control Valve	This component controls the flow of coolant into the heater core. When the temperature knob on the climate control head is moved to hot, the heater control valve opens to allow coolant to flow through the heater core. When the temperature know is moved to cool, the heater control valve closes to prevent coolant from flowing into the heater core. Some vehicles do not use this device.

Fuel Systems

The fuel system will be covered in two parts. First, the induction system will be covered. This system manages, measures, and directs air flow into the engine. Next, the fuel components of a fuel-injection system will be covered.

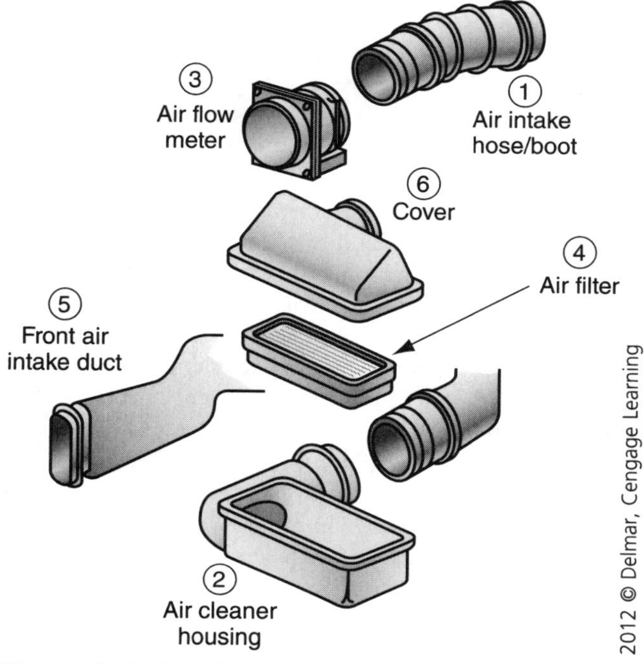

Figure 3. Induction System

Induction System

Diagram Number	Component Name	What It Does
1	Air Intake Hose/ Boot	This hose connects the air cleaner to the throttle body in most fuel-injected vehicles. It may have connections to the positive crankcase ventilation (PCV) system. The hose provides the flex between the body and the engine; thus, a break may cause strange drivability problems due to false, uncalculated air entering the system.
2	Air Cleaner Housing	The housing holds the air filter. There are as many different configurations as there are vehicle models, but they all hold the air filter. Some manufacturers mount the air meter or mass air flow sensor on the air cleaner, as in the illustration. Only filtered, clean air can flow past the air flow meter.
3	Air Flow Meter	here are two types of meters commonly in use: the *vane* meter and the *mass air flow* meter. The vane-type meter has a moving flap in the incoming air tract that opens and closes based on how much air the engine is using. The

(Continued)

Induction System

Diagram Number	Component Name	What It Does
		computer uses this signal to calculate how long to open each injector. The other type of meter uses a heated wire across the intake stream. When air crosses this wire or grid, the wire cools off. This change in temperature is used by the computer to calculate how much air the engine is using and how much fuel to deliver.
4	Air Filter	The *air filter* element has the job of removing dirt and other contaminants from the incoming air to the engine. All of the various designs perform this same function. These are usually replaced at recommended mileage intervals. An extreme mileage interval may require more frequent change.
5	Front Air Intake Duct	The front air intake duct is where the air that enters the engine is picked up. This component is the top part of the air cleaner housing. It is necessary to remove this device when servicing the air filter.

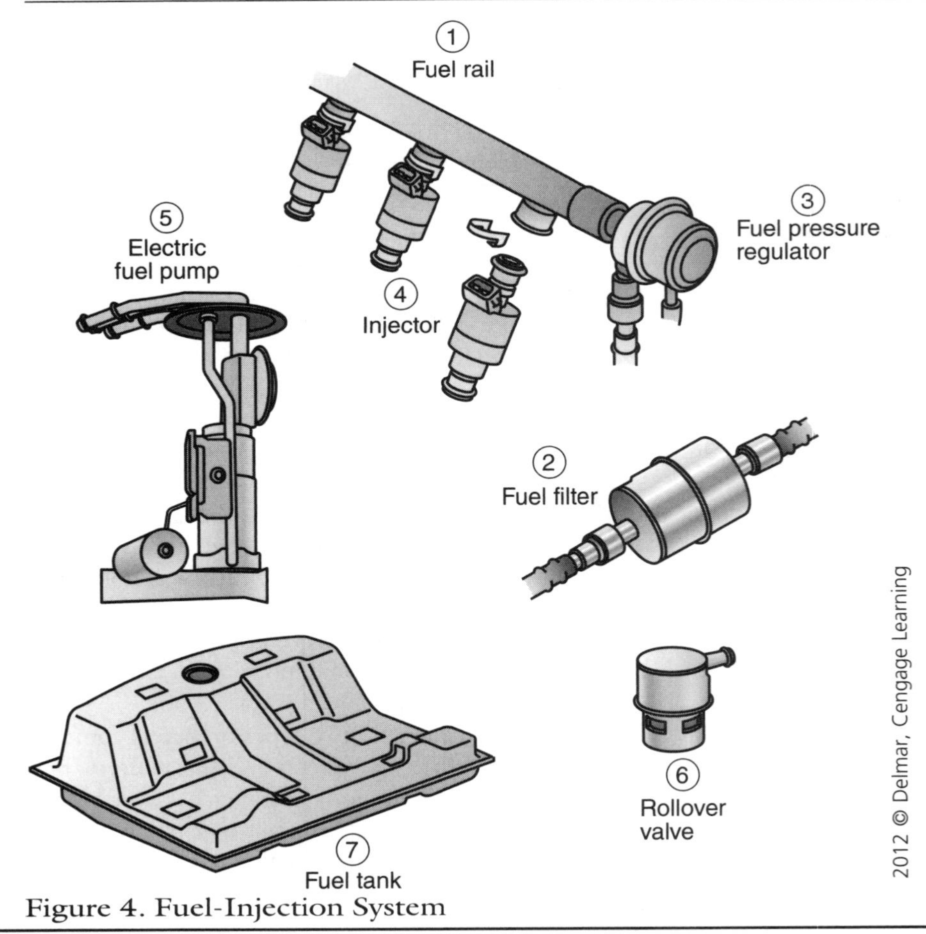

Figure 4. Fuel-Injection System

Fuel Injection System		
Diagram Number	**Component Name**	**What It Does**
1	Fuel Rail	The *fuel rail* connects the fuel lines to the fuel injectors. In many applications, the fuel pressure regulator is attached to the fuel rail. The fuel injectors are retained in the fuel rail by o-rings and, in some applications, by clips. This is typically not a part that wears out.
2	Fuel Filter	*Fuel filters* contain the filtering media used to remove dirt and debris from the fuel system after the fuel pump, but before the fuel injectors.
3	Fuel Pressure Regulator	The *fuel pressure regulator* is located in the engine compartment most of the time. It resides in the return side of the fuel line and keeps the fuel pressure at the necessary level by restricting the amount of fuel returned to the tank. The control for the pressure may be a fixed spring inside the regulator or vacuum from the engine. In a returnless system, fuel pressure may be controlled by the power train control module (PCM). These systems do not have a return line; instead, they vary the speed of the electric fuel pump. In this situation, the regulator is a monitoring device telling the computer how much pressure is being produced.
4	Fuel Injector	A *fuel injector* is an electrical component. It is a high-speed solenoid. When the computer completes the electrical circuit to the *pintle*, which is a small needle valve, it is lifted up off the seat and fuel sprays out of the tip in a nice cone shape. When the computer releases the circuit, the valve closes. This is measured by the technician in milliseconds. An injector is typically open for 1–2 milliseconds at idle and 12–15 milliseconds under heavy load. This rapid movement and the cycle from cold to hot and back again are the most common causes of injector failure.
5	Fuel Pump Module	The *fuel pump module* houses the fuel pump and, in many cases, the fuel gauge sending unit, along with any plumbing necessary to complete the fuel return to the tank. Most manufacturers sell this as an assembly. Some fuel pumps are sold separately and are installed in the original module or mounted in other places on the vehicle. The in-tank unit is, by far, the most common design.

(Continued)

Fuel Injection System		
Diagram Number	**Component Name**	**What It Does**
6	Rollover Valve	The rollover valve serves two purposes. The first is to keep fuel from running out of the fuel tank vent line in the event of a vehicle rollover. The second is a check valve for the evaporative-emission system. Evaporative-emission systems in fuel-injected vehicles are really designed to control fuel vapors that might leave the tank as the engine warms the fuel from the constant circulation through the fuel rail and back to the tank. In vehicles built since 1996, the evaporative system is controlled and monitored by the engine-management computer or PCM. The fuel tank is a critical and major component of this system. The largest evaporative system line is usually connected to the rollover valve and then to the engine on the other end. A solenoid-controlled vacuum valve is used by the PCM to apply vacuum to the tank for testing and to draw vapor into the engine to be burned.
7	Fuel Tank	In late-model vehicles, the *fuel tank* is more than just a holding compartment for fuel. It functions as a surge tank to manage the changes in fuel temperature. It usually contains the fuel pump and a sending unit. It also has baffles or partitions itself inside to help the fuel stay as close to the fuel pump pick up as possible.
8	Strainer	Prevents large debris from being drawn into the fuel pump and likely being pushed forward to the fuel filter.

Ignition System

The ignition system is an electrical system. It is made up of two parts. The one most obvious to our customers is the part that includes the *ignition switch*. This part of the ignition system has grown from a simple power switch for the vehicle engine to the master control switch for the entire vehicle. The second part of the ignition system is what technicians have traditionally called the *ignition system*. This includes the components that have been the tune-up parts. Ignition components provide the high-voltage, low-amperage spark to fire the combustible gases to start. With so many changes in vehicle maintenance over the last decade, the tune-up has become a very gray area that does not really fit anymore. The components of both a conventional distributor electronic ignition and distributorless ignition system will be covered in the following section.

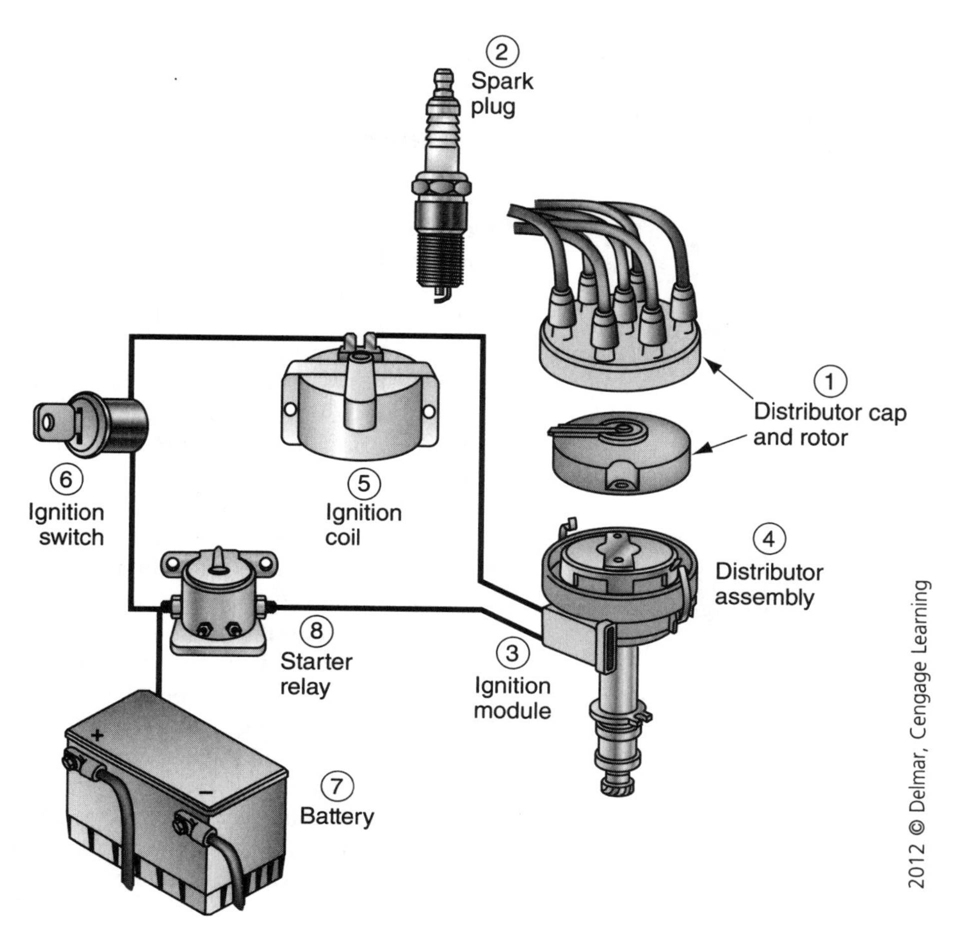

Figure 5. Electronic Ignition System with Distributor

Electronic Ignition System with Distributor

Diagram Number	Component Name	What It Does
1	Distributor Cap and Rotor	Probably the most familiar component (with the exception of the spark plug), the *distributor cap* has a small button in the center that carries current from the ignition coil to the center electrode of the rotor. The *rotor* is attached to a shaft in the distributor that turns at the same speed as the engine's valve train. When the tip of the rotor passes under one of the contacts in the distributor cap, the electricity it is carrying finds a ground through the spark plug at the other end of an ignition wire that is attached to the plug and the cap. Due to the very high voltage levels that jump across these air-capped gapped connections, the cap and rotor are subject to deterioration over time.
2	Spark Plug	*Spark plugs* are very simple devices. They provide a ground for the ignition system through the threaded body that bolts into the cylinder head.

(Continued)

Electronic Ignition System with Distributor		
Diagram Number	**Component Name**	**What It Does**
		When current is applied to them, a small electrical spark jumps across the inner electrode, which is insulated from the outer shell to the outer electrode attached to the spark plug shell. This spark starts a chemical reaction in the combustion chamber that causes the fuel and air mixture to oxidize very rapidly, heating it and causing expansion. Spark plugs may have electrodes made of, or coated with, copper, platinum, or titanium to help them last longer. When spark plugs wear, the electrodes that are very square to the eye start to round and erode. This wear will cause the ignition system to work harder to make that spark jump the gap.
3	Ignition Module	The *ignition module* receives a low-voltage pulsed signal from either the distributor pick-up coil, the PCM, or the crankshaft position sensor. The ignition module then amplifies the pulsed signal of 5 volts or less to a 12-volt signal that is sent to the coil(s). The ignition module amplifies the signal again to several thousand volts while firing the spark plugs. In computer-controlled applications, the ignition module usually acts as middle management between the PCM and the ignition coil. In this role, it takes commands from the PCM for timing and often provides the fixed or "base" timing necessary to start the engine. It then keeps the engine running until the computer takes over the responsibility.
4	Distributor	The *distributor* has a mechanical connection to the engine and turns at the speed of the valve train, one-half crankshaft speed. In earlier vehicles that had a mechanical spark advance, the distributor had a vacuum-controlled advance mechanism as well as centrifugal advance. Late-model vehicles that are equipped with distributors use computer-controlled timing.
5	Ignition Coil	The *ignition coil* takes a low-voltage pulsing signal from the ignition module, amplifies it from about 9–12 volts to 5,000–40,000 volts and sends it to the spark plugs. Coils may be mounted remotely using an attaching high-tension wire. They may be integral or directly mounted to the spark plug, referred to as a coil over plug (COP) design.
6	Ignition Switch	The ignition switch is used to give the driver a method of starting the engine.
8	Starter Relay	The starter relay is used on some vehicles to send a start signal to the starter. The starter relay receives the signal from the ignition switch.

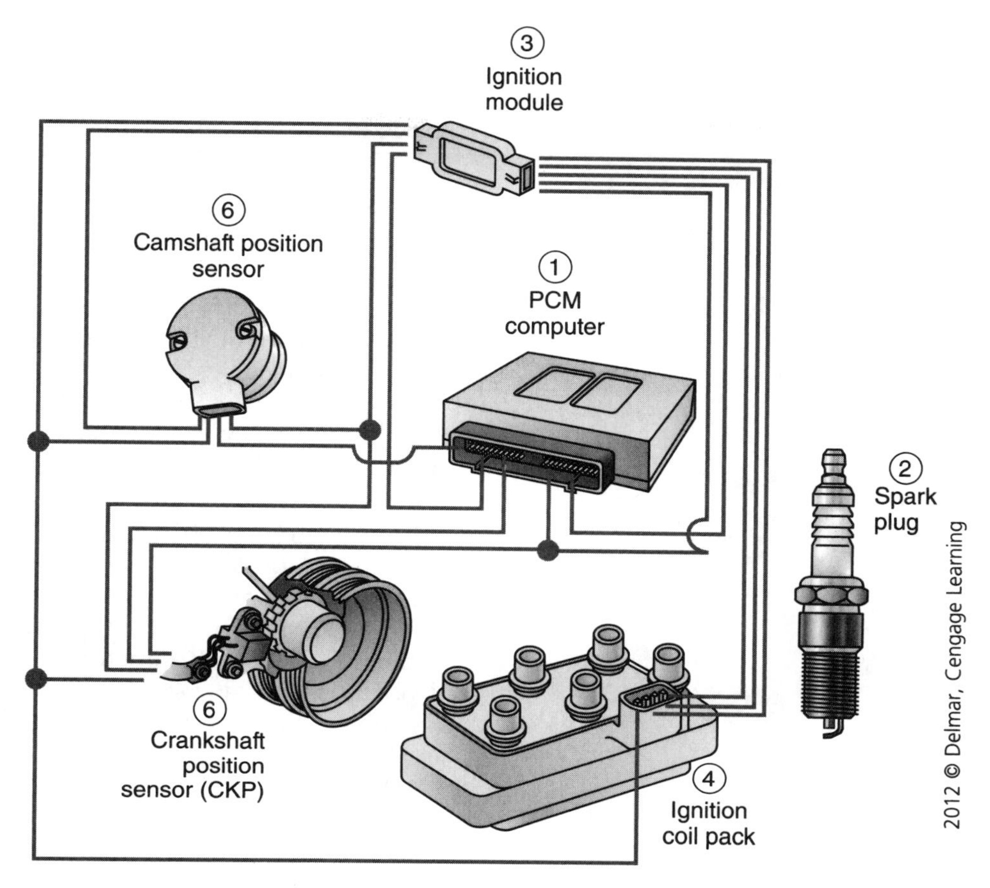

Figure 6. Electronic Distributorless Ignition System

Electronic Distributorless Ignition System

Diagram Number	Component Name	What It Does
1	PCM (Computer)	The *power train control module* (PCM) is the engine control unit. The PCM takes inputs from various sensors and operates various actuators. There is an extensive description of PCM operation in the emission section. For the purposes of ignition control, the PCM manages ignition timing by varying a signal between the ignition control module and itself. If the vehicle is equipped with knock sensors, it will control timing when the engine experiences spark knock or pinging, usually by retarding timing.
2	Spark Plug	Spark plugs are very simple devices. They provide a ground for the ignition system through the threaded body that bolts into the cylinder head. When current is applied to them, a small electrical spark jumps across the inner electrode, which is insulated from the outer shell to the

(Continued)

Electronic Distributorless Ignition System		
Diagram Number	**Component Name**	**What It Does**
		outer electrode attached to the spark plug shell. This spark starts a chemical reaction in the combustion chamber that causes the fuel and air mixture to oxidize very rapidly, heating it and causing expansion. Spark plugs may have electrodes made of, or coated with, copper, platinum, or titanium to help them last longer. When spark plugs wear, the electrodes that are very square to the eye start to round and erode. This wear will cause the ignition system to work harder to make that spark jump the gap.
3	Ignition Module	The ignition module receives a low-voltage pulsed signal from either the distributor pick-up coil, the PCM, or the crankshaft position sensor. The ignition module then amplifies the pulsed signal of 5 volts or less to a 12-volt signal that is sent to the coil(s). The ignition module amplifies the signal again to several thousand volts while firing the spark plugs. In computer-controlled applications, the ignition module usually acts as middle management between the PCM and the ignition coil. In this role, it takes commands from the PCM for timing and often provides the fixed or "base" timing necessary to start the engine. It then keeps the engine running until the computer takes over the responsibility.
4	Ignition Coil Pack	The ignition coil takes a low-voltage pulsing signal from the ignition module, amplifies it from about 9–12 volts to 5,000–40,000 volts and sends it to the spark plugs. In distributorless ignition systems, there may be multiple coils. In some applications, cylinders share a coil, while coil-on-plug systems have a coil for each spark plug. These connect directly to the spark plug and are referred to as a coil over plug (COP) design.
5	Crank Position Sensor	In the distributorless system, the PCM or computer must have a way to determine what position the crank is in rotation so that it knows when to fire the spark plugs. This is accomplished by using a degreed wheel on the crank or flywheel with a special signature that tells the PCM which location is cylinder #1. This helps the PCM synchronize the spark plugs to the crankshaft position. Because the crank makes two revolutions for each revolution of the valve train, spark plugs are fired twice during a complete cycle, once during the power stroke and once during the exhaust stroke. This is referred to as *waste spark design*.

Electronic Distributorless Ignition System		
Diagram Number	**Component Name**	**What It Does**
6	Cam Position Sensor	In some systems, an additional sensor called the *camshaft position sensor* takes the place of the special signature on the crank sensor wheel. This special signature is the reference to start the sequence of ignition firing in the correct order to provide a start. This is used to determine the exact position of the camshaft and cylinder #1. This added sensor is most commonly found in systems that use sequential multiport fuel injection, when the PCM needs to fire both the injector and the spark plugs at precise times in reference to crank position.

Exhaust System

Since it is apparent that exhaust systems are the plumbing coming out of the engine, this section will be brief. The nomenclature and descriptions of the components are listed below.

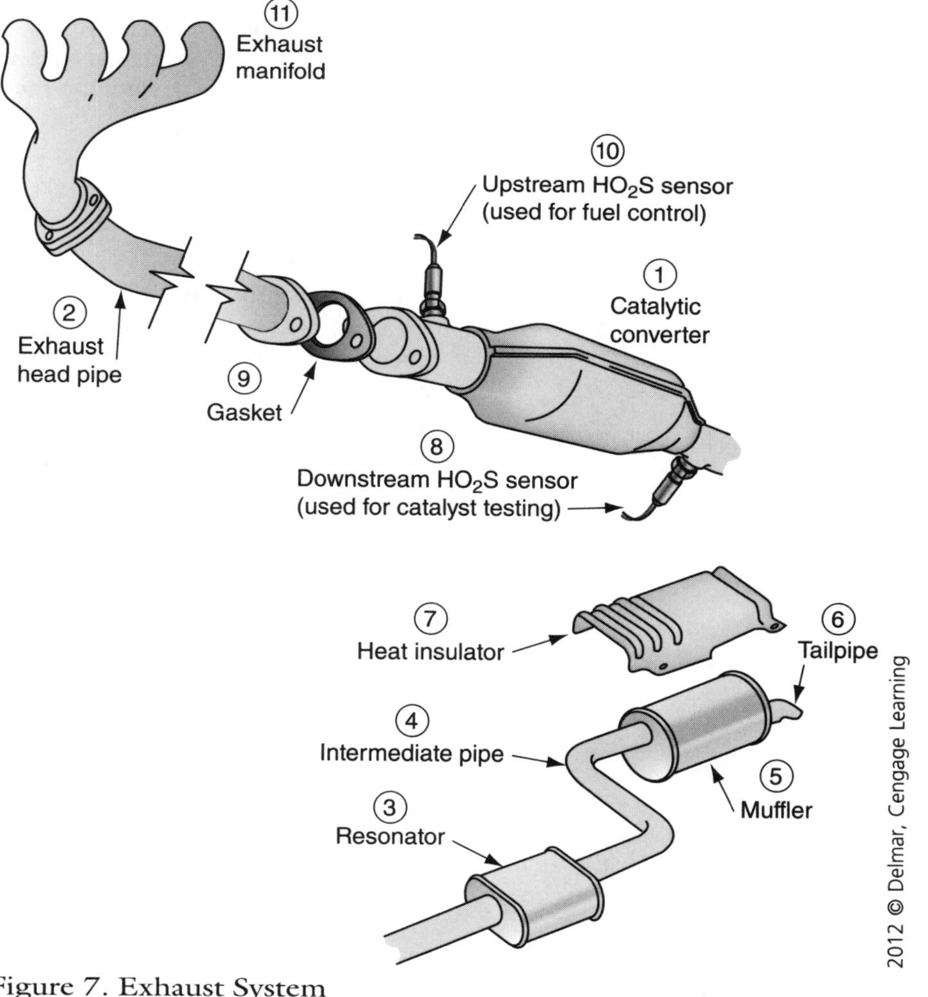

Figure 7. Exhaust System

Exhaust System		
Diagram Number	**Component Name**	**What It Does**
1	Catalytic Converter	The most complex component of the exhaust system is an emission component with no moving parts. The *catalytic converter* is located at the front of the vehicle, usually as far forward as possible. The catalytic converter has the job of removing or reducing several by-products of combustion. These components are carbon monoxide (an odorless, invisible, and deadly gas), hydrocarbons (small quantities of unburned fuel), and NOx or oxides of nitrogen (not to be mistaken for nitrous oxide), which are responsible for photochemical pollution (acid rain). These emissions are removed by passing the exhaust stream over a bed of precious metals that cause a chemical reaction with the exhaust when they are hot. The conversion is designed to create carbon dioxide and water from the combustion by-products. This is why water can be seen as droplets from the tailpipes of late-model vehicles and as steam upon the warm-up of the engine.
2	Exhaust Head Pipe	The exhaust head pipe is the pipe directly behind the exhaust manifolds on the engine. It often has the catalytic converter built in.
3	Resonator	Some exhaust systems have a resonator in them to help remove some of the engine's noise. They may be located before or after the muffler, depending on what the engineers were trying to accomplish.
4	Intermediate Pipe	The pipes between the head pipe and the muffler are known as *intermediate pipes.* On V-engines with dual exhaust, sometimes there is a pipe that joins the two sides of the engine in the intermediate pipe. This pipe is called a *crossover* or *H-pipe.*
5	Muffler	The *muffler* is a resonating chamber that cancels out loud noises from the engine. There is usually a perforated tube that runs through the muffler. The perforations allow sound to be reflected to the chambers within the muffler. They are sometimes filled with damping material like steel wool or ceramic panels.
6	Tailpipe	Just as it sounds, this is the pipe that exits the exhaust out from under the vehicle at its rear or side.

Exhaust System		
Diagram Number	Component Name	What It Does
8, 10	Oxygen Sensors	These sensors monitor gases passing through the exhaust piping. They provide critical input for emission control.

Emission Control Systems

Emission control systems encompass nearly every engine control system. In addition to understanding the evaporative system in the fuel system section and the catalytic converter in the exhaust, the service consultant can gain knowledge of the basics of emission control systems by looking at a general computer-controlled system and some of its possible emission control devices.

Because different vehicles require different types of equipment, there will not be any specific system questions here, but great benefits come from understanding how the systems work. These can be very expensive components to repair and they seldom have much effect on the way the vehicle runs. This is why the service consultant must be able to help the customer understand the implications of driving a vehicle with the ***malfunction indicator light*** (MIL) on. The MIL lamp is amber in color to indicate caution.

The modern engine management system, or power train control module (PCM), is an amazingly adaptable and dependable system. It makes vehicles start and run much better and more consistently than pre-computer-controlled vehicles. It is surprising to know that the original reason for having on-board computers was actually to control emissions. The easiest way to understand these systems is to relate them to our own bodies. The computer is like our brain. It has a vast store of information and "rules" programmed into it. It can adapt to changes in weather, driving style, road conditions, and even vehicle wear. Unlike the humans who designed them, PCMs do not have to make many mistakes to learn, and they can learn in milliseconds. That does not mean that when they get poor inputs they will not make bad decisions. When they do, the service consultant is the front line "psychologist" to help the PCM get back on the right path.

So, how do these little boxes come to all these rapid and accurate decisions? One easy way to think of it is to understand that they use electronics to replicate human senses. For example, smell: The computer uses one or more oxygen sensors mounted in the exhaust system to sniff the exhaust stream to see if it is delivering too much or too little fuel. It can even use this information to confirm a misfire condition or a bad catalytic converter. In reality, its sense of "smell" is just an electrical signal (usually, 0.1–0.9 volts or 100–900 millivolts) sent to the PCM and called an ***input***.

Moving on to feel or touch, the computer uses several sensors to this end. Coolant- and air-temperature sensors communicate the temperature. The mass air flow sensor feels the amount of air the engines breathing and informs the PCM. A crank and cam sensor help the PCM monitor the engine's speed and track misfires. Many engines use a set of electronic ears to listen for pinging or detonation; this set is thus labeled a ***detonation sensor***. The PCM uses this information to adjust timing and protect the engine from damage. The PCM "sees" the position of the driver's foot on the throttle by tracking voltage through a variable resistor, called a ***throttle-position sensor***. There are many other inputs to the PCM. The shop's technicians can help provide an understanding of these inputs when diagnosis and repair require it.

The PCM evaluates all of these inputs and responds to provide the best emissions and drivability to the driver. PCM responses include varying timing and fuel delivery many times per second; tracking inputs like air conditioner commands and power steering inputs to adjust idle speeds; and responding to load to vary the application of emission control devices like air pumps, evaporative systems, and exhaust gas recirculation (EGR) valves.

The *exhaust gas recirculation system* (EGR) is used to reintroduce a controlled amount of exhaust back into the combustion process. Because exhaust is basically an inert gas that will not create any heat during combustion, it helps to cool the process down under light loads. This serves two purposes. First, it lowers NOx emissions; second, it helps to cool down the internal parts of the top end of the engine like cylinder heads, pistons, and valves. Cooling them can help eliminate the pinging that can occur under leaner fuel conditions. Pinging or detonation occurs on the compression stroke when the air and fuel mixture fires before the piston is all the way up. Because the piston is connected to the crank, it must keep going up; the force created by the false ignition process, however, makes a knocking or pinging sound as the flame front hits the top of the piston like a hammer. When this happens, the engine is misfiring and creating large amounts of all types of emissions. Prolonged detonation can severely damage the piston and/or engine.

Air pump systems have been in use since 1969, when some of the manufacturers used them to get their high-performance engines past the California emissions standards. This system pumps ambient air into the exhaust stream to help finish combustion of any burnable by-products in the exhaust system. Air injection is necessary for some types of catalytic converters to be efficient. The system is made up of a belt-driven or electric pump on the engine, lots of plumbing, and some one-way check valves to keep exhaust from getting into the pump. These systems are controlled by components ranging from simple vacuum valves to complex interfaces built between the PCM and the air management valve. These systems allowed manufacturers to continue using older engine designs that were not as efficient and still maintain lower emission standards. Most vehicles no longer use air pumps, as newer engine designs use more efficient means to lower emissions.

Starting/Charging Systems

This section will not have pictures because it is typically not necessary to know about every component of the cranking or charging systems. Below is a brief explanation of how the cranking and charging systems function.

When the driver wants to start the engine, he/she begins by turning the key that sends power to the starter. Some manufacturers use a starter relay in conjunction with the ignition/start switch. The starting circuits are the highest amperage circuits in the vehicle and may be easily identified by their large cables. These include the battery cables. When all of the electrical connections are made, the *starter* (cranking motor) begins to spin, pushing a starter drive gear into the flywheel drive gear at the same time. Once the engine begins to run, the starter drive backs away from the flywheel, because the engine is running faster than the drive. The driver releases the key, and the starting system becomes dormant until the next start up. Always remember that a cranking motor, many times referred to as a starter, *only* provides rotation for starting. Other requirements are needed for an engine to run. The most common causes of starter failure are failure to engage due to open internal circuits or high resistance between circuit connections.

The battery is simply a storage device. It stores the necessary electrical energy to start the vehicle, and it provides a reserve when loads exceed the alternator's output. Battery cables are the most frequently overlooked cause of repeat failure of alternators and batteries.

The *alternator* (in the past and recently referred to again as the *generator*) and *voltage regulator* are the charging system components. Most modern alternators have integral voltage regulators, which means that the voltage regulator is inside the alternator. Automotive electrical systems use *direct current (DC)* voltage. Alternators perform their work in *alternating current (AC)* voltage. This AC voltage is converted to DC voltage inside the alternator through the use of diodes. The voltage is then controlled by the voltage regulator. According to Ohm's law, the more load that is on an alternator, the less voltage it will produce. The opposite is also true. The voltage regulator controls voltage, attempting to maintain around 14.3 volts. Alternators are generally serviced as complete units, so the cause of failure is often summarized as the inability to meet specifications for amperage loads or minimum voltage. Alternators are very hard working components in late-model vehicles. Some are capable of generating as much as 130 amps. Failure occurs due to wear and heat, in most cases.

2. Drive Train Systems (3 Questions) (Includes manual transmission/transaxles, automatic transmission/ transaxles, and drive train components)

Manual transmissions and transaxles are used on some late-model vehicles. These units require the use of a third pedal in the floor, the *clutch* pedal. The service consultant must be able to drive these types of vehicles in order to properly communicate with customers who own them. In addition, service consultants will likely have to move these vehicles in and out of the repair shop, as well as take test drives with some customers.

Manual Transmission/Transaxles

The manual transmission uses a clutch and manual shifting to change gears. The service consultant will not need to make diagnostic decisions in the C1 test. Diagnosis should always be left to the technician. The service consultant will only need to know how the system functions and how to identify major components. The figure here shows a transaxle. Item number 4 discusses the differential, which is used on transaxles. Transaxles are used on front-wheel drive vehices. These units usually contain a differential gear set which connects to the front drive axles.

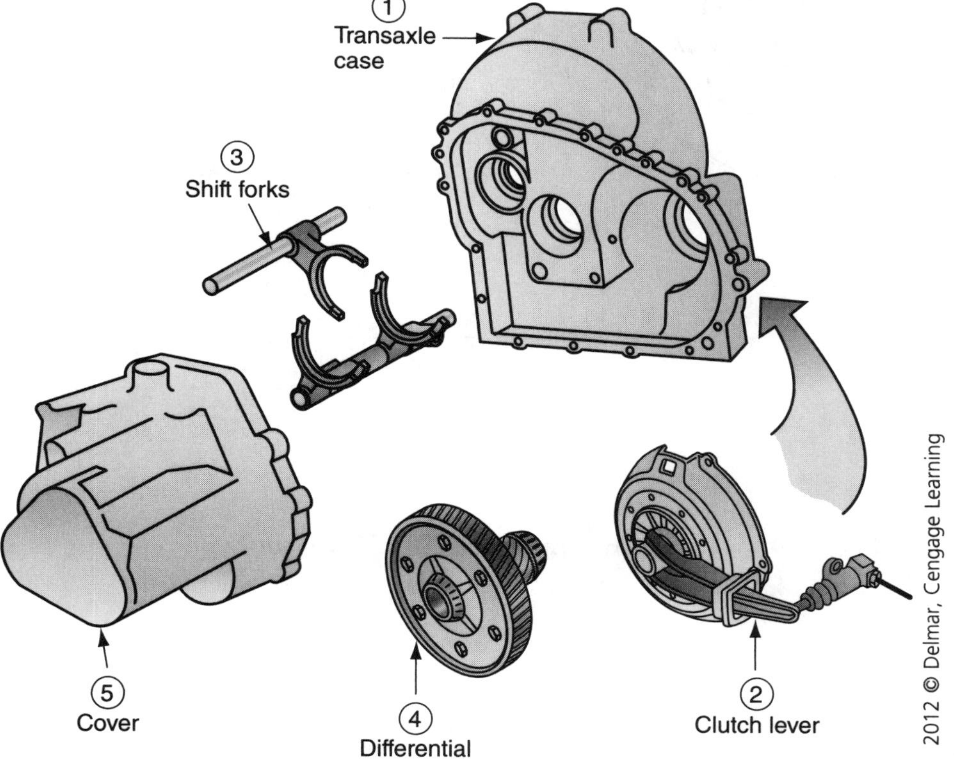

Figure 8. Manual Transmission/Transaxles

Manual Transmission/Transaxles		
Diagram Number	**Component Name**	**What It Does**
1	Transaxle Case	All of the components attach to, or are housed by, this assembly.
2	Clutch Lever	The *clutch lever* is the external connection to either the hydraulic cylinder or cable that operates the clutch inside the bell housing.
3	Shift Forks	Manual transmissions use *shift forks* that straddle the synchro/gear assemblies to move the synchro from gear to gear, causing a shift.
4	Differential	In a front-wheel drive, the *differential* is usually inside the case and driven by a final drive gear from the transmission gear train that engages the large ring gear.
5	Cover	An access point is available on some transaxles. The final drive gears are often in this cover.

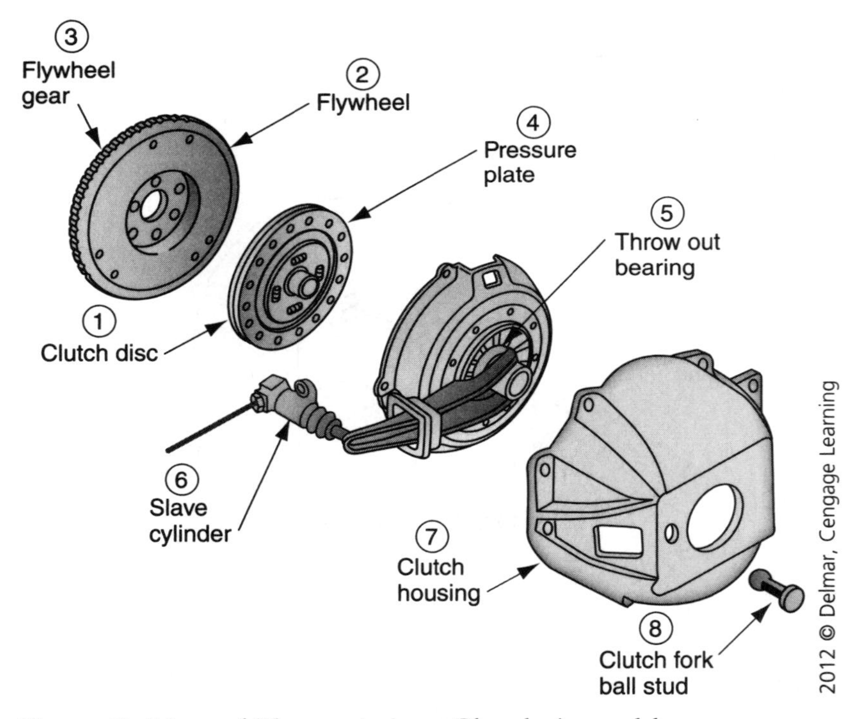

Figure 9. Manual Transmission: Clutch Assembly

Manual Transmission: Clutch Assembly

Diagram Number	Component Name	What It Does
1	Clutch Disc	The *clutch disc* is the friction part of the clutch. The springs in the center cushion the clutch engagement. The *clutch disc* is two-sided and riveted together. The disc slides onto and drives the transmission input shaft.
2	Flywheel	The *flywheel* provides one side of the disc-clamping surface. It is machined smooth before assembly and bolts to the rear of the crankshaft.
3	Flywheel (Ring) Gear	This gear is pressed onto the flywheel and is the gear the starter engages when turning the engine over during starting.
4	Pressure Plate	The *pressure plate* provides the clamping force and is actuated by the clutch fork to release the clutch.
5	Throw Out Bearing	The *throw out bearing* is a sealed bearing that rides against the pressure plate, actuating fingers during disengagement. Many late-model vehicles apply light pressure so that the throw out bearing is in constant contact with the pressure plate.
6	Slave Cylinder	Some clutches use a cable to connect the pedal assembly to the clutch actuation fork. Others use a hydraulic system similar to brakes. The *slave cylinder* may be inside or outside the bell housing. The slave cylinder expands to push on the clutch fork when the pedal is depressed, and the clutch master cylinder generates hydraulic pressure. This action and relationship is similar to a brake master and wheel cylinder.
7	Clutch Housing	The clutch housing provides a place to mount the clutch slave cylinder or mechanical linkage.
8	Clutch Fork Ball Stud	The clutch fork ball stud serves as the pivot point for the clutch lever.

Automatic Transmission/Transaxles

The modern automatic transmission is a fairly complex unit. Just like the manual transmission, it provides different gearing to help the engine move the vehicle. Unlike the manual transmission, an automatic transmission uses circular bands that operate by holding a member of a planetary gear set than can create different gear ratios. The automatic transmission is controlled hydraulically. Instead of a clutch that must be engaged and disengaged, it uses a hydraulic coupling device, known as a torque converter, that varies the amount of engine torque and engagement. The internal controls of the transmission are fed hydraulic pressure by pumps inside the transmission and controlled by the valve body that directs shifts by changing pressures and directing fluid through passages. Below is a diagram of some of the major components of the automatic transmission system, along with a chart of their names and purposes.

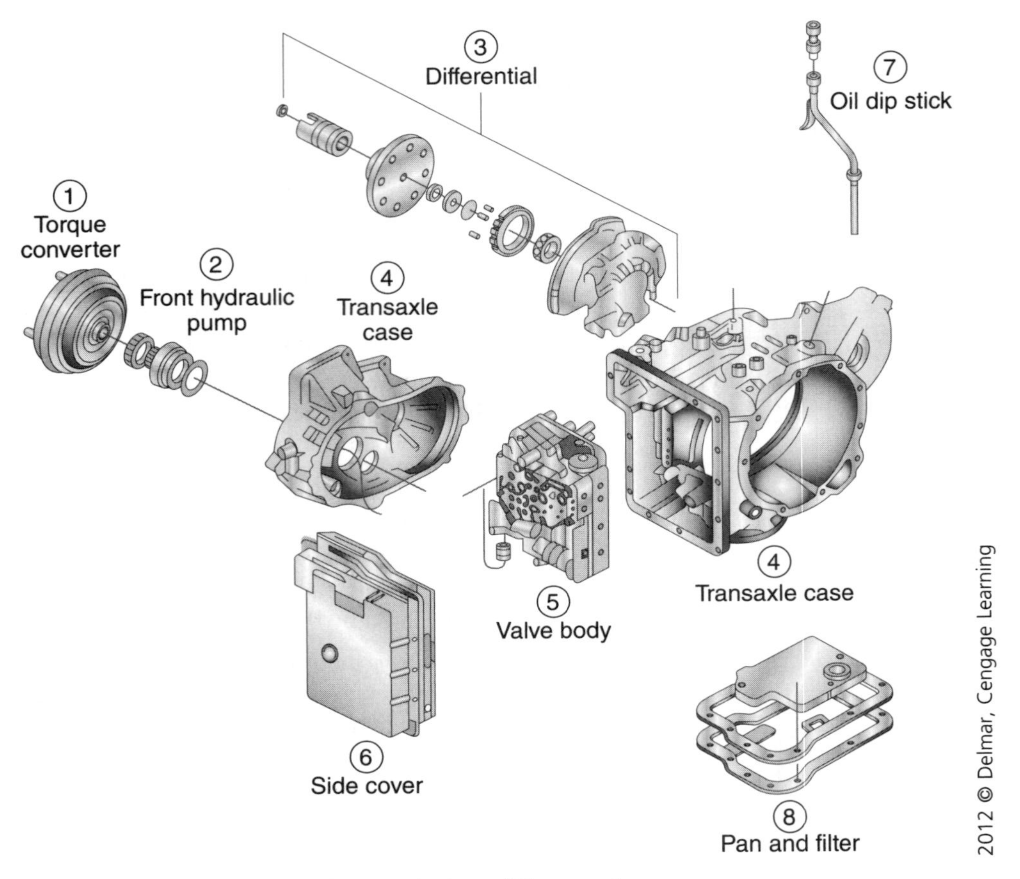

Figure 10. Automatic Transmission/Transaxles

Automatic Transmission/Transaxles

Diagram Number	Component Name	What It Does
1	Torque Converter	The *torque converter* is a hydraulic coupling device that connects the engine to the transmission. Most late-model automatics have an electric locking clutch inside the converter that makes a direct connection between the engine and the transmission during light load cruising. This electric locking clutch drops RPM, helps with gas mileage, and is referred to as a lock-up torque converter. This design now raises the miles per gallon (MPG) used by an engine with an automatic transmission to be the same as, or similar to, one with a manual transmission.
2	Front Hydraulic Pump	This is one of the primary pumps in the transmission. Few transmissions have a rear pump as well.
3	Differential	In a front-wheel drive vehicle, the *differential* is usually inside the case and driven by a final drive gear from the transmission gear train that engages the large ring gear.

Automatic Transmission/Transaxles

Diagram Number	Component Name	What It Does
4	Transaxle Case	All of the components attach to, or are housed by, this assembly.
5	Valve Body	The hydraulic control unit. Newer electronic transmissions have solenoids within the *valve body* that are commanded by the PCM.
6	Side Cover	Usually a cover for the valve body in front-wheel drive applications. Most rear-wheel drive vehicles locate their valve body in the bottom of the transmission in the oil pan area.
7	Oil Dip Stick	Most automatic transmissions have a fluid-level *dip stick*, while most manual gearboxes do not.
8	Pan and Filter	The pan is the storage location for the transmission fluid. Most late-model vehicles have either a screen or a felt filter that is usually serviceable by removing the pan.
Not Pictured	Planetary	The *planetary* is a multiple-ratio gear set used in automatic transmissions to provide different gear ratios in a compact package. It is called a planetary because it is a system of a single gear with multiple gears that revolve around it, similar to the way the planets revolve around the sun.

Drive Train Components

Drive train components include the rest of the components that connect the engine and transmission to the wheels. The components include driveshafts, constant velocity (CV) axles, differentials, and 4-wheel drive or all-wheel drive transfer cases. There are many designs, so looking at generic versions will promote understanding about how they work and interrelate.

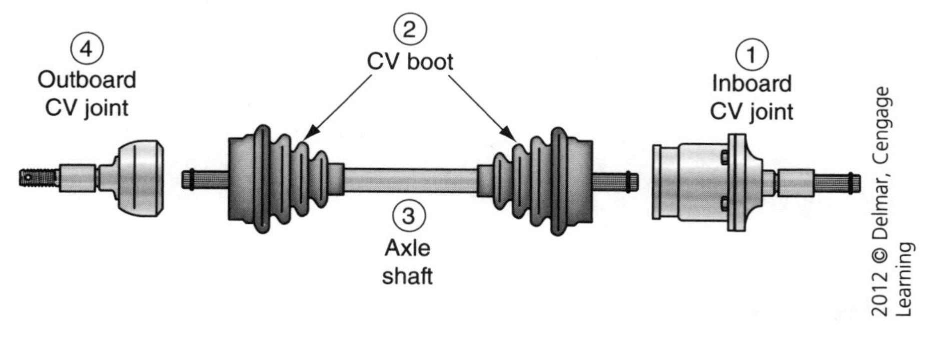

Figure 11. Drive Train Components

Driveshafts: CV Axle		
Diagram Number	**Component Name**	**What It Does**
1	Inboard Constant Velocity (CV) Joint	The inner *constant velocity* (CV) *joint* snaps into the differential side gears. This joint only operates in two planes. It must handle in and out movement, called *plunge*, as the vehicle moves up and down, changing the length of the driveshaft. It must also handle moving up and down with the suspension.
2	CV Boot	The major maintenance item on CV axles, the *boot* holds the lubricant in the joint and has bellows-shaped ridges that help it move with changes in position and keep contaminates, like dirt, water, etc., out. The outer boots have the highest failure rate due to all the movement they must provide.
3	Axle Shaft	The *axle shaft* has splined areas at each end to engage and mount both CV joints. It may be hollow or solid steel.
4	Outboard CV Joint	The workhorse of the CV axle, this joint must operate in multiple planes at once while rotating. The outer joint must turn with the wheels and move up and down with the suspension. CV axles are used in front-wheel drive vehicles to both wheels, on front axles in late-model 4-wheel drives with independent suspension, on rear differential applications on mini 4-wheel drives, and even on some driveshaft applications.

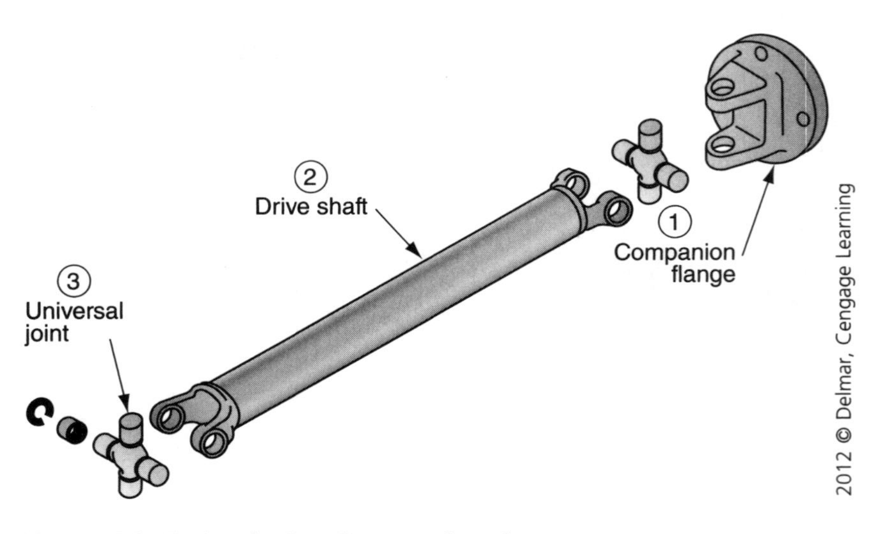

Figure 12. Driveshafts: Conventional

Driveshafts: Conventional		
Diagram Number	**Component Name**	**What It Does**
1	Companion Flange	This type of flange bolts to a differential or transfer case flange. Many vehicles will place the actual U-joint bearing cups directly into a yoke on the differential or transfer case output and retain them with a U-bolt. This is becoming the standard approach because of the repeatability of installation, which helps to control vibration.
2	Driveshaft	The *driveshaft* provides the connection between the transmission or transfer case and the differential. They may be constructed of solid or hollow steel tubing, aluminum, or carbon fiber. In most applications, the U-joints are pressed into the driveshaft and retained by clips.
3	Universal Joints (U-joints)	*Universal joints* (U-joints) allow the driveshaft to make a connection while rotating and moving up and down with the suspension. Four cups hold needle bearings that allow the U-joints to turn to accommodate the movement.

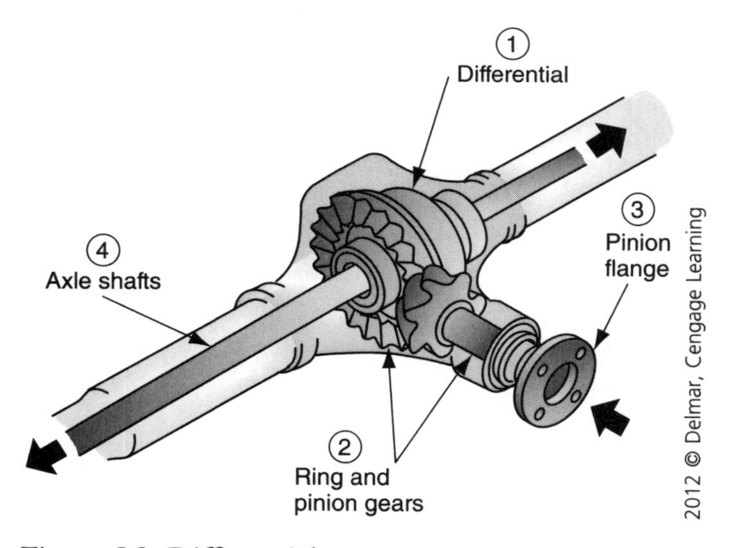

Figure 13. Differential

Differential		
Diagram Number	**Component Name**	**What It Does**
1	Differential	Differentials may be found at either the front or rear of the vehicle. Front-wheel drive vehicles incorporate the differential into the transaxle. Four-wheel drive vehicles use a differential at each end of the vehicle. Rear-wheel drives have one at the rear of the vehicle to assist in turning.
2	Ring and Pinion Gears	In a conventional rear-wheel or 4-wheel drive vehicle, the driveshaft connects to one of two large gears called the *pinion gear*. The gear that it meshes with changes the direction of the driveshaft rotation 90 degrees. It is called the *ring gear*. The ring gear is bolted to the carrier and causes the carrier to turn. In front-wheel drives the engine's crank runs parallel to the drive axles. There is a form of final drive from the transmission part of the transaxle that drives the ring gear and carrier assembly.
3	Pinion Flange	The *pinion flange* provides a means to connect or couple the driving shaft to the rear assembly.
4	Axle Shafts	These shafts, usually made of solid steel, rotate at different speeds and provide connection to the wheel and tires.

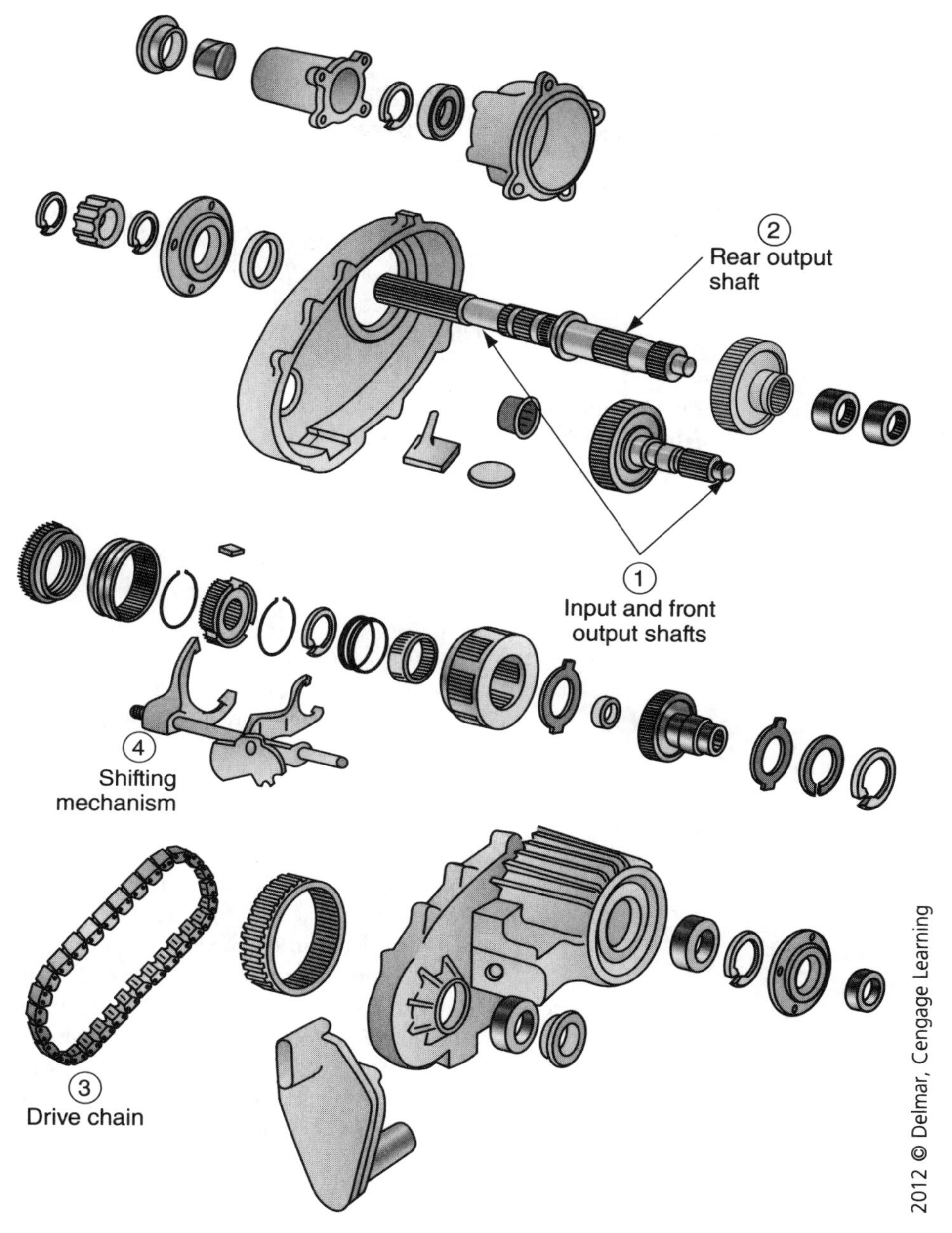

Figure 14. 4-Wheel Drive Transfer Case

A *transfer case* is a gearbox that engages or disengages the front and rear axles. There are usually two different ratios inside the box, which will be discussed further.

4-Wheel Drive Transfer Case		
Diagram Number	**Component Name**	**What It Does**
1	Input and Output Shafts	The transfer case is either mounted behind or physically bolted to the rear of the transmission via the *input shaft*. The front *output shaft* connects to the front driveshaft and the front axle/differential. The rear output shaft connects to the rear driveshaft.
2	Rear Output Shaft	The *rear output shaft* connects to the rear driveshaft and rear axle/differential. This is the shaft that is used in both 2- and 4-wheel drive applications. Unless the transfer case is shifted to neutral, this driveshaft always is driven by the engine and transmission, providing vehicle movement.
3	Drive Chain	Many transfer cases use a chain to drive the front output shaft from the main shaft when 4-wheel drive is engaged. Other transfer cases use a gear-to-gear front drive, which is very heavy duty in construction.
4	Shifting Mechanism	Conventional transfer cases have a shift lever inside the vehicle to allow the drive to select between 2-wheel drive (rear wheels only), 4-wheel low (all four wheels driven slower while the engine runs faster for climbing hills or doing heavy work), and 4-wheel high (all four wheels driven at normal speed). Some transfer cases have a *synchronized shift* to allow them to be shifted "on the fly." Newer systems have electronics that use a motor to change the gear selection instead of a shifter. There are many variations, but these are the most common.

3. Chassis Systems (3 Questions) (Includes Frames, Brakes, Suspension, Steering, Wheels, Tires, and TPMS)

Frame

The frame acts as the vehicle's skeleton to which other chassis components are attached.

Brakes

It is important to spend some time looking at the brake system because it is a system in which service consultants must be well-versed. The diagram and chart display a typical disc/drum system and a generic antilock brake system.

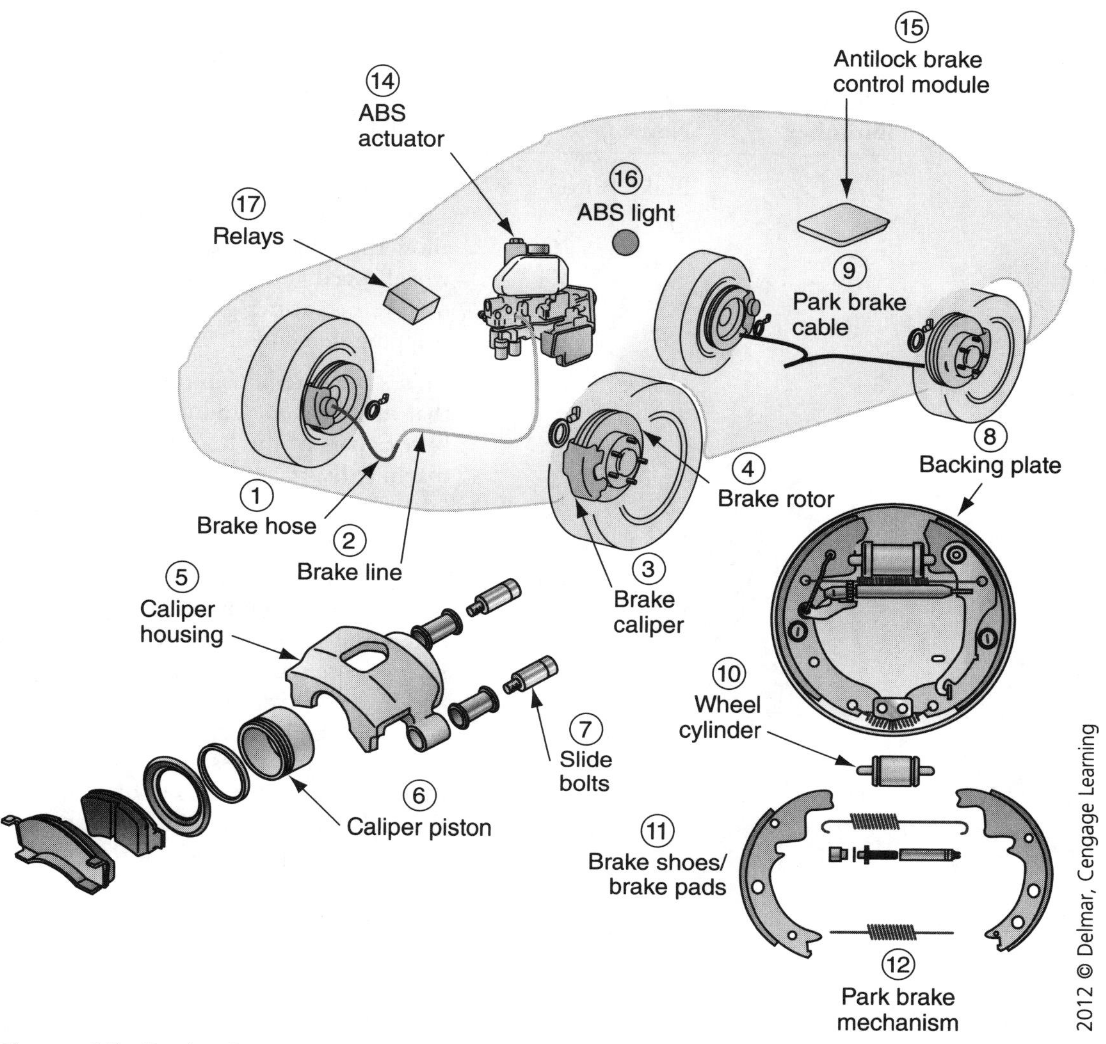

Figure 15. Brake System

Brake System		
Diagram Number	**Component Name**	**What It Does**
1	Brake Hose	Flexible hoses that connect the brake components at the wheels to the hard brake lines on the body. They allow suspension movement.
2	Brake Line	Nonflexible hard lines that are attached to the body of the vehicle. They run from the various components, starting at the master cylinder and working out to where the brake hoses connect to them.
3	Brake Caliper	The *brake caliper* is the moving component of a disc brake. It works like a clamp to grab the spinning rotor and slow the vehicle.

(Continued)

Brake System		
Diagram Number	**Component Name**	**What It Does**
4	Brake Rotor	This is the namesake of the disc brake system. The *brake rotor* has two machined sides that allow the caliper to squeeze. The rotor rotates at wheel speed and also acts as a heat sink to shed high temperatures generated by friction due to stopping demands.
5	Caliper Housing	This is the bare housing component of the caliper that functions as a mounting bracket. The piston bore is part of the housing and often has machined surfaces that allow the assembly to "float" on pins or slides to compensate for wear of the brake pads.
6	Caliper Piston	The piston is like a deep, round puck. It may be made of steel, aluminum, or special plastic compounds. Most brake calipers have only one *caliper piston*, but high-performance applications and heavy vehicles may have multiple pistons. When the brake pedal is depressed, the master cylinder creates pressure in the system, which causes the piston to push out of its bore toward the brake rotor. The opposing force of the brake pad on the opposite side of the rotor causes a clamping force that stops the vehicle.
7	Slide Bolts	Many brake systems use bolts that tighten into the spindle. The caliper floats on them. Others may use plates that allow the caliper to slide. These slides are critical for even brake pad wear. If the caliper does not slide, the inner pad often wears out very quickly because the pressure on the rotor is not equal on both sides.
8	Backing Plate	On drum brake systems, the *backing plate* is the mounting plate for all of the components of the system.
9	Park Brake Cable	The *park brake cable* connects at one end to the parking brake pedal or lever. At the other end, it attaches to the drum brake by a special cam mechanism that pushes the brake shoes out mechanically to secure the vehicle in a parked position. Rear-disc applications use a piston-locking mechanism to hold the brake in park.
10	Wheel Cylinder	The *wheel cylinder* performs the same purpose as the brake caliper. Instead of clamping, however, it pushes out on the brake shoes.
11	Brake Shoes/ Brake Pads	The *brake shoe* or *disc pad* has a very complex job. Designed to wear out as it does its job, these parts are made up of metallic and organic materials that

Brake System		
Diagram Number	**Component Name**	**What It Does**
		are called friction materials. These components work by riding next to either a rotor or drum. When the brake pedal is pushed, the hydraulic system functions to move the brake shoes/pads into the drum/rotor to create friction.
12	Park Brake (Emergency Brake) Mechanism	If a hydraulic failure were to occur, a backup mechanical system would be of help. The *park brake mechanism* mechanically actuates the rear brakes in all but a very few applications. When there are disc brakes on the rear, a similar device mechanically pushes the pads out to clamp the rotor. This handy safety device is almost always linked to the components that automatically adjust the brakes. Using the park brake helps to adjust the rear brakes. Backing up the vehicle will also actuate the self-adjusters of most drum brakes.
13	Master Cylinder	The *master cylinder* is the first component of the hydraulic portion of the brake system. When the brake pedal is depressed, it generates pressure to actuate the brake calipers and wheel cylinders. Pressure can be as high as 2,000 psi.
Not Pictured	Power Brake Booster	*Brake boosters* provide assistance to push the brake pedal in power brakes. Most use vacuum applied to the master cylinder side of a chamber with a diaphragm in the center to supply this assist. Some systems are known as hydro-boost and use a similar arrangement with the power steering pump, supplying hydraulic pressure for assist.
Not Pictured	Pedal Assembly	The *pedal* is the component that attaches to the master cylinder pushrod and gives the driver a mechanical advantage due to a pedal ratio designed into the mechanism. This advantage can result in 10 pounds of pedal pressure creating 100 pounds of brake pressure without other assistance, similar to the advantage created by a lever.
Not Pictured	Combination/ Proportioning Valve	In vehicles not equipped with an *antilock brake system (ABS),* the *proportioning valve* is used to control the amount of pressure sent to the rear brakes. If pressures are equal at all four wheels, the rear brakes will lock up first because of the difference in weight. Most vehicles are lighter in the rear than the front. Also, in stopping forward momentum, weight transfers to the front brakes. This component usually has a pressure

(Continued)

Brake System		
Diagram Number	**Component Name**	**What It Does**
		differential valve/switch that will shut down part of the brake system should a large loss of pressure occur, as with a leak. It will also turn on the red brake light on the dash.
14	ABS Actuator	An electro-hydraulic component in ABS vehicles, the *ABS actuator* contains the electronic versions of the proportioning valves that are run by the ABS control module to control wheel lock-up.
15	ABS Control Module	This component analyzes inputs from the wheel speed sensors and sometimes the vehicle speed sensor to control wheel lock-up. This is very much a simplification because of the vast number of systems of varying complexity out there.
Not Pictured	Brake Warning Lamp	The red *brake warning lamp* comes on when the park brake is not fully released, the brake fluid is low, or the pressure differential switch in the combo valve has tripped. In some applications, the ABS system can turn this light on along with its malfunction indicator light. Remember, red lights have higher significance than amber.
16	ABS Warning Lamp	This light comes on when the ABS system is not functioning. Some systems turn it on when brake fluid is low. When both the ABS and the MIL lamps are on or flashing, it indicates immediate danger or component damage.

Suspension

The suspension is what holds the vehicle up, provides a smooth ride, keeps the tires in contact with the ground, and provides safe control when turning. The following section will review each of the components in a short, long-arm (SLA) suspension, which is the most common type. The version selected uses a suspension strut and is typical of both front and rear suspensions of many vehicles. Before beginning, two other types of suspensions commonly in use will be covered.

The leaf spring suspension is found on the rear of many vehicles and the front of heavy duty pick-ups. In this suspension an axle housing is mounted with a leaf spring on each side of the vehicle. The spring functions as a control arm, locating the rear end fore and aft in the vehicle, in addition to providing normal spring function. More and more vehicles, including trucks, are being designed with independent suspension on all four wheels, so this design will probably only remain on heavy trucks in the future.

The second variant is the McPherson strut. In this design the strut functions as the upper suspension attaching point. These suspensions do not have an upper control arm. The strut provides a pivot point at the top in the form of a strut bearing or mount that allows the strut to turn with the wheels. This design also provides much better handling than riding

SLA designs. The strut pictured in the diagram is not a McPherson strut; it is not a suspension member. It functions as the shock with an integral coil spring. The SLA suspension gets its name from the short upper control arm and long lower control arm that make up its design. This design handles very well because it allows the tire to move in the direction of negative camber as the body rolls going around corners. Camber will be discussed along with alignment at the end of this section. These newer independent designs provide better tire-to-road contact. This not only promotes long tire wear, but adds safety due to advanced tire friction-to-road contact or connection.

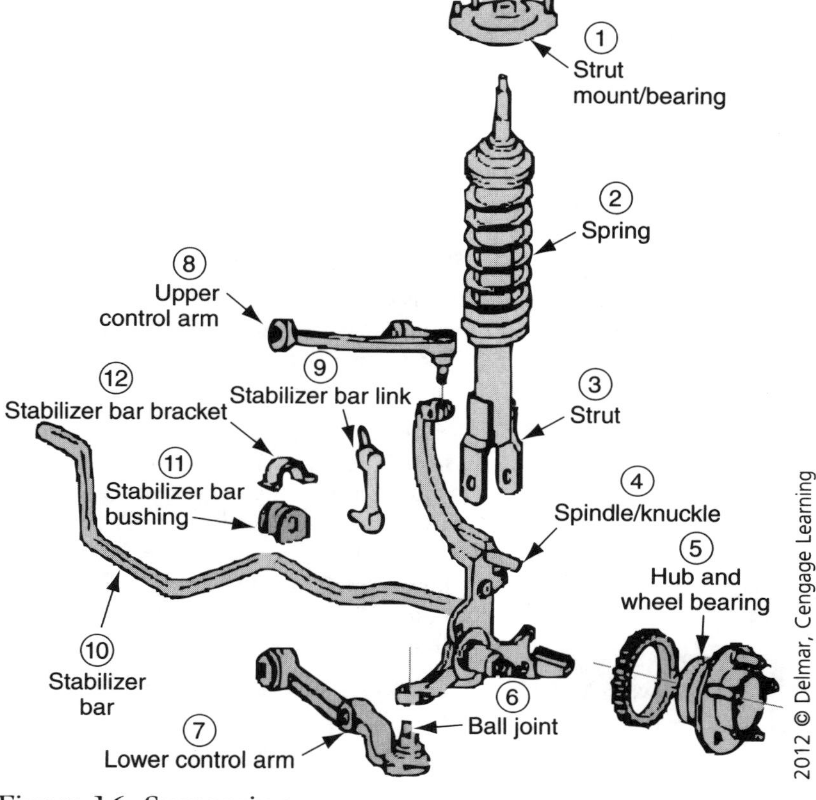

Figure 16. Suspension

Suspension

Diagram Number	Component Name	What It Does
1	Strut Mount/ Bearing	The *strut mount* cushions and isolates the strut assembly from the body. In McPherson applications, it also houses a bearing to allow the strut body to rotate when the wheels turn.
2	Spring	The *coil spring* is most common. The previously mentioned leaf spring is used with live axle applications, and one other variation known as the *torsion bar* round out the spring types used in modern vehicles. The torsion bar is a long straight spring that functions by twisting. It is

(Continued)

Suspension		
Diagram Number	**Component Name**	**What It Does**
		attached at one end to the body and the other end to the suspension. It is usually adjustable to attain correct preload and ride height.
3	Strut	The *strut* is a shock absorber that functions as a mount for the coil spring in some applications. The SLA suspension may use a strut like the diagram, or it may have a shock absorber.
4	Spindle/Knuckle	The steering *knuckle* or *spindle* provides the movement during steering along with the mounting areas that accommodate the wheel hubs, brake components and the ball joint ends of the control arms.
5	Hub and Wheel Bearing	This component is shown as a stand-alone piece, but it may be integrated into a brake rotor with serviceable bearings. The type shown uses a sealed bearing.
6	Ball Joint	*Ball joints* are like the joints in our shoulders or hips. They provide for the multi-plane movement that occurs as the vehicle moves up and down and when the wheels turn. They get their name by a ball-shaped bearing surface that rides in a lined cup. The ball joint is usually bolted or pressed into the control arm and has a long stud that is bolted into the steering knuckle.
7	Lower Control Arm	The *lower control arm* is the longer one of the two. It joins the suspension from the steering knuckle to the body of the vehicle. It often has mounting points for shocks, stabilizer bars, and sometimes the springs.
8	Upper Control Arm	The *upper control arm* is the short arm. It joins the suspension from the steering knuckle to the body of the vehicle. It, too, often has mounting points for shocks springs.
9	Stabilizer Bar Link	There are several designs, but the *stabilizer bar link* or *end link* connects the ends of the stabilizer bar to the suspension.
10	Stabilizer Bar	Also known as a *sway bar*, the *stabilizer bar* is an example of a torsion bar. When a force is applied to one side of it, there will be an equal and opposite force on the other end. Thus, as a vehicle rolls to the outside as it makes a turn, the force on the outside wheel will be transferred to the inside wheel to keep the tire in contact with the ground. This is sometimes referred to as an *anti-roll bar*.

Suspension		
Diagram Number	**Component Name**	**What It Does**
11	Stabilizer Bar Bushing	To allow the stabilizer bar to pivot on the body of the vehicle, a bushing is located on the body equal distances from the center.
12	Stabilizer Bar Bracket	This is the bracket that holds the sway bar and bushing to the body of the vehicle.

Steering

Only two types of steering discussed here: *rack and pinion* and the *linkage style*. Some recent systems are electrically driven by motors using varying input voltage to a control module. These systems will not have a power steering pump since they use electrical assist. The main thing to remember with these systems is that they are diagnosed with a scan tool and sometimes have to be calibrated after adjustments are made to the steering system.

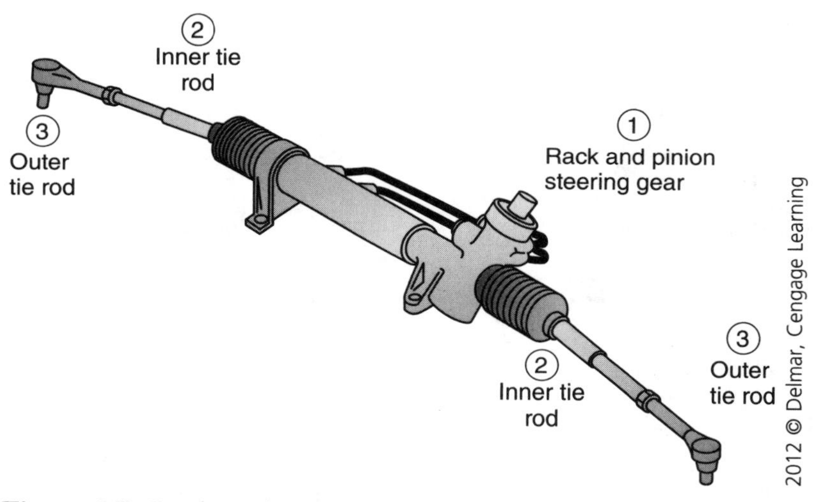

Figure 17. Rack and Pinion

Rack and Pinion		
Diagram Number	**Component Name**	**What It Does**
1	Rack and Pinion	*Rack and pinion steering* gets its name from the internal components. The *rack* is a long gear that runs horizontally inside the housing. The tie rods attach to the ends of the rack. The *pinion* is a small gear that connects to the steering column shaft. The reason for using gears is to change the ratio between the wheel's turn and steering input. These can be a manual design, though most are power-assisted. When a rack and pinion is power-assisted, a valve is located at the top of the pinion that provides hydraulic assist when there is

(Continued)

Rack and Pinion		
Diagram Number	**Component Name**	**What It Does**
		movement of the pinion; fluid lines connect to the unit.
2	Inner Tie Rod	*Tie rods* are like small versions of ball joints: They allow rotational and multi-plane movement to follow suspension movement. The inner socket on the rack and pinion is an open joint because a bellows boot similar to a CV joint boot covers it and protects it from contamination.
3	Outer Tie Rod	The outer tie rod connects the inner tie rod to the steering knuckle. The two thread together and allow alignment changes to the toe setting by adjusting their lengths. The use of left- and right-hand threads allows for this. Compare replacements with the original for correct thread pitch and match.

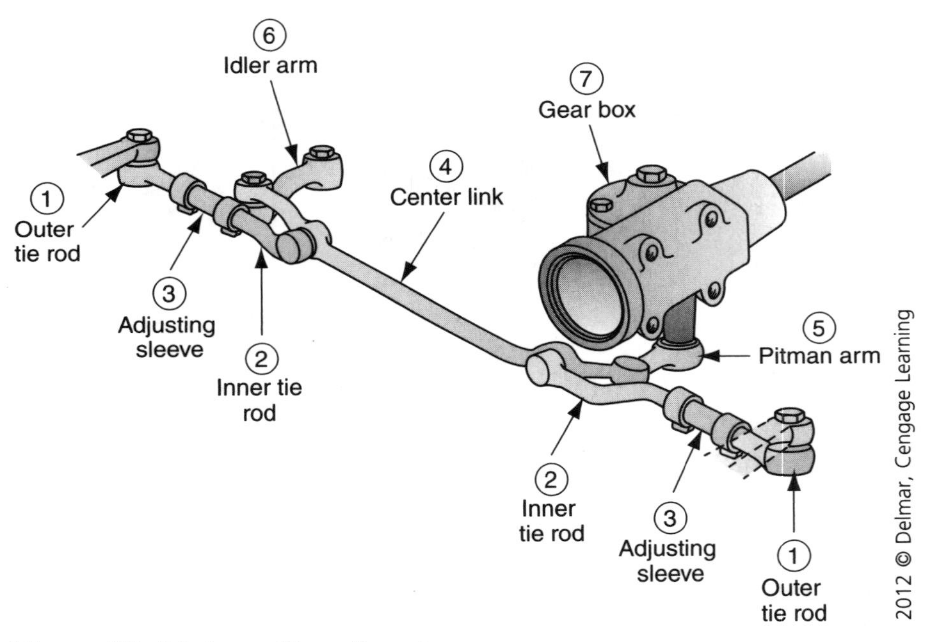

Figure 18. Linkage-Type Steering

Linkage-Type Steering		
Diagram Number	**Component Name**	**What It Does**
1	Outer Tie Rod	Tie rods are like small versions of ball joints that allow rotational and multi-plane movement to follow suspension movement. The outer tie rod connects the inner tie rod to the steering knuckle.

Linkage-Type Steering		
Diagram Number	**Component Name**	**What It Does**
		The two thread together and allow alignment changes to the toe setting by adjusting their lengths. The components of the suspension must move up and down at the wheels. The components of the steering must allow this movement while causing the least amount of change in the toe settings of the alignment. Tie rod ends allow around 30 to 60 degrees of movement. Looseness due to wear or binding is a reason for replacement.
2	Inner Tie Rod	The inner tie rod works just like the outer tie rod. In most cases, there is an *adjusting sleeve* between the two. The inner tie rods may attach to a center link, the tie rod for the other side of the vehicle, or directly to the steering box. Many inner and outer tie rods are threaded with one left- and the other right-handed so that when the technician adjusts them, they rotate in the same direction to adjust toe-in or -out on each side of the vehicle.
3	Adjusting Sleeve	The *adjusting sleeves* are generally made of steel and have clamps at each end where the tie rods thread into them. After the technician adjusts the toe to specification, he tightens the clamps on the adjusting sleeve to lock the toe setting. When one or both tie rods are replaced, the adjusting sleeves are often rusted and frozen, necessitating their replacement as well. Often, they are made with right- and left-handed threads, so it is important to assemble and install them correctly.
4	Center Link	The *center link* is almost always used in conjunction with an idler arm. The center link attaches between the steering box and idler arm and maintains a parallel plane across the front suspension. In most center link applications, the tie rods attach to the center link and then to the steering knuckles. A similar variation of this theme is the *drag link*. The drag link generally attaches from the steering gear to a spindle. Like the center link, it does not provide for the adjustment, but may be adjustable to allow the steering box or wheel to be centered.
5	Pitman Arm	The *Pitman arm* is the large lever that bolts to the bottom of the steering output shaft. Some may have ball-like joints that attach them to the rest of steering linkage. These require replacement when they wear. The other type of Pitman has a hole to

(Continued)

Linkage-Type Steering		
Diagram Number	**Component Name**	**What It Does**
		receive the joint of a tie rod and does not usually have to be replaced.
6	Idler Arm	The *idler arm* is located on the passenger side of the vehicle. In all but a few applications, there is one idler arm. The idler arm is responsible for providing mirror image movement of the idler arm or its opposing Pitman arm to keep the center link in precise parallel alignment. The bushings in the idler arm are the parts that generally wear resulting in a wandering sensation on the road.
7	Steering Gear	The *steering gear* may be power-assisted or manual. It has a gear arrangement using a set of ball bearings that run in a cage between the gears to provide smooth steering operation and minimum reaction back to the driver when going over bumps. This design is known as *recirculating ball steering*. It is the most prevalent design in use in the last 30 years. Common failures are usually leaks at output- or input-shaft seals and leaking power-steering line connections. The mechanical part of steering boxes is highly dependable due to the importance of its function. These systems are bulky and heavy compared to rack and pinion setups. They work best in heavy-duty applications for these reasons.

Wheels, Tires, and TPMS

As with most of these areas, wheels and tires can be a whole specialty unto themselves. Questions within this test will be limited to knowledge of tire and wheel service issues and not engineering. A few of the common conditions that occur with wheels and tires are listed below.

When a tire is out of round it is called *radial run out*. This will appear like high or low spots as the tire turns around. This can cause road vibrations, if it is bad enough, even after multiple attempts to balance the assembly.

Another condition that occurs with defective tires, bent wheels, or tires that have had a blow to the sidewall is called *lateral run out*. Tires with this condition may appear to move side to side when spun on a balancer. The vehicle may demonstrate a vibration at speed and a sort of side-to-side wiggle when moving slowly.

Wheel balance and rotation are maintenance items for tires. *Wheel balance* is the process of spinning the wheel on the balancer. The balancer looks for the heaviest spot on the wheel and calculates the necessary weight to put on the spot directly across from it to equalize the assembly's dynamic weight. This is a severe over-simplification, but is more than enough for the basic understanding needed. The balancer uses both the inside and the outside of the wheel to work out a dynamic balance. Newest balancers apply road force to best provide solutions to the imbalance.

Wheels can contribute to these problems if they have been bent or are damaged from road hazards or improper installation. Most late-model steel wheels are somewhat flexible and can be damaged very easily. If the lug nuts that hold the wheels on are not properly tightened, a wheel can distort, causing a problem that may feel like a tire problem.

Many late-model vehicles are equipped with a tire pressure monitoring system (TPMS). These systems alert the driver by illuminating an indicator in the instrument panel when low tire pressure is present in any of the vehicle tires. These systems are very sensitive and will cause the warning indicator to illuminate if the pressure drops just 2 or 3 psi lower than the specified setting. Special testers are available to diagnose problems in this system.

The ability to explain the alignment angles to a customer is very valuable. Here are the basic angles that may show up in questions on the test.

- **Camber:** When a person looks at the wheel from the front, the amount the tire leans in or out at the top is called *camber*. When the tire leans out it is called *positive camber,* and when it leans in at the top it is called *negative camber.* Camber is a wear angle, which means that if it is out of spec, it can cause the tires to wear on the shoulder to the direction the top of the tire leans. Out at the top causes outside shoulder wear, in at the top causes inside shoulder wear.
- **Caster:** *Caster* is the forward or rearward tilting of the steering axis as viewed from the side of the vehicle. Caster is not a wear angle. Excessive caster can cause a vehicle to steer heavily in parking lots or even make the steering wheel shake going over bumps. Vehicles with too little caster will wander and follow ruts in the road, causing an oversteer. In years past, cars without power-assisted steering would be aligned with much negative caster to assist in turning at slow speeds.
- **Toe:** *Toe* is the direction the front of the wheel points when going down the road. If a person stands up and points his or her toes toward each other, that is what is known as *toe-in.* If a person points them away from each other, that is known as *toe-out.* Vehicles with excessive toe-in will wear the tires across the face. They may not exhibit any unusual driving tendencies because toe-in is the more stable of the two settings. Vehicles with excessive toe-out will wear the tread across the face in the opposite direction. The vehicle may exhibit wander or be somewhat unstable when applying the brakes, produce squeal, or rapidly wear.
- **Thrust Angle:** *Thrust angle is the relationship between the overall toe direction of the rear wheels to the overall toe direction* of the centerline of the chassis. If both rear wheels are pointed to one direction or the other, the front steering must compensate to make the car go straight. Vehicles that have a thrust line problem can often be spotted when following them. The person following can see the side of the vehicle from directly behind it, but not the other side. The body appears to be going crooked down the road. Modern alignment equipment takes this angle into account when setting front toe to keep all four wheels going in the same direction. Readings are taken between all wheels through precise measuring instruments.
- **4-Wheel Alignment/2-Wheel Alignment:** Most late-model vehicles that are front-wheel drive and many rear-wheel drive vehicles require a 4-wheel alignment. This means that sensors are mounted on all four wheels so that all of the wheels can be set. Traditional 2-wheel alignment meant that only two sensors were put on the front of the vehicle with none on the rear. This method is not practiced anymore due to the ability to use a thrust line on a rear-wheel drive vehicle. Even when only two wheels have adjustment, all four sensors are mounted to allow this compensation and to provide a means to look for unforeseen damage or bent components.

4. *Body Systems (3 Questions) (includes Body Components, Glass, Heating and Air Conditioning, Electrical, Restraint, and Accessories)*

Body Components

Repairs that need to be made on the vehicle body usually will need to be sent to a automotive body shop. It is not uncommon for a mechanical repair shop to sublet a body repair to a nearby body shop. See Task C.2 for more information on sublet repairs.

Glass

Repairs that need to be made on the vehicle glass are usually performed by a body shop or by a glass specialty shop. Either of these options would be classified as a sublet repair. See Task C.2 for more information on sublet repairs.

Heating and Air Conditioning

The heating and air conditioning (A/C) systems provide comfortable air inside the cabin of the vehicle. Warm air is generated using the heat created by the engine. The conductor for this is known as the *heater core*. The heater core is a small radiator. Coolant from the engine is circulated through it, and a blower blows air across the core to provide hot air inside the vehicle. There are several passages for the air to flow through that are controlled by the controls on the dash. The controls may have cables, vacuum actuators, or electric motors that run a series of doors inside the ducts to direct the air where the driver wants it. Temperature is controlled by a blend door, and in some cases, by a water control valve that limits the amount of warm engine coolant that enters the heater core.

Air conditioning works through the same set of ducts. The A/C system uses a compound that has a very low boiling point. The refrigerant at this time is R-134A. The A/C system takes advantage of the refrigerant's low boiling point to move it through two states—liquid and gas.

An air conditioner has a compressor that provides both suction and discharge. It creates movement and pressure within the system. The compressor pushes gaseous refrigerant to the condenser, which sits in front of the radiator. The refrigerant enters the condenser as a high pressure gas. As the refrigerant flows through the condenser, the refrigerant changes back to a high pressure liquid. After leaving the condenser, the refrigerant is directed to a metering device that is located near the evaporator core. This liquid is then evaporated back to gas state. Only gas state refrigerant can pass through a compressor; liquid will not compress.

A/C systems use either a *thermal expansion valve* or an *orifice tube* to create a restriction. This causes the liquid refrigerant to spray into the evaporator. At this point, the pressure on the liquid drops, and it rapidly becomes a gas. This process is called *evaporation*. This causes the evaporator to become very cold so that a fan blowing across it will yield cold air. The compressor starts the whole cycle over again. The components of the A/C system share most of the ducting with the heater, but have electrical controls to engage the compressor. The system may have switches to protect the compressor from extremely low or high pressures. Many A/C systems are monitored and controlled by a body control computer or the PCM.

Electrical

Current automotive electrical systems are 12-volt DC, and they use the negative battery terminal as ground. There are many body systems that use electricity to operate. They include power windows, door locks, cruise control, radios, lighting, seats, doors, anti-theft systems, suspension controls, engine cooling fans, heating, and A/C. There is virtually no system on a modern car that is not controlled or monitored by an electrical device. This is why the condition and the capability of the charging system are so critical. The word *electrical* generally encompasses to electronic, that is, solid-state controls without moving parts, such as, transistors, diodes, transformers, etc.

Restraint

Restraints refer to the safety items inside the vehicle. Seat belts and air bags are the most common items. Vehicles are equipped with all kinds of reminder and warning systems that encourage passengers to use the restraint systems.

The airbag or *supplemental restraint system* has a couple of key components that need to be noted. The airbag modules themselves are the devices that rapidly deploy when a crash sensor detects that a blow to the vehicle might cause an injury the airbag could lessen or prevent. These modules are one-time use and must be replaced when they deploy. The other item that should be noted is the clock spring or spiral cable. The *clock spring* is mounted under the steering wheel, and it allows the steering wheel to rotate while maintaining a hard-wire connection to the airbag inflator. This component also provides an electrical path for all other steering-wheel-mounted electrical controls.

Seat belts are the other restraint area that may come up on the test. Modern seat belt designs allow the passenger to move their upper body freely to maneuver. They have a load-sensing inertia device in them that locks the belt under potential emergency or accident avoidance maneuvers. Some seat belts are equipped with pre-tensioners that fire off like airbags to keep the occupant in place during an accident. Child seat safety anchors were not included in the task list at this time.

Accessories

Accessories are the last area under body systems. Questions in this section will typically be very broad and general in nature. Test takers with general knowledge of how basic mechanical and electrical systems work and interact with each other should be able to answer questions in this section. The practice questions give several examples from this area.

5. Services/Maintenance Intervals (3 Questions)

1. Understand the elements of a maintenance procedure.

There are two different paths that can be taken when maintaining a vehicle. The first is following a maintenance schedule that performs standard items and services based on the wear characteristics of the vehicle. The second path is damage-control repair upon failure. It has been proven in many studies that the maintenance path is the most cost-effective approach. To understand and explain a maintenance procedure, the service consultant must know what the procedure is trying to accomplish. Here is an example: Most manufacturers recommend that the fluid in the transmission be changed every 25,000–30,000 miles. The fluid is exchanged for new fluid and, in some applications, a filter is replaced as well. It is important to know what the steps are and why they are performed in order to explain it to the customer. Some maintenance may also be needed to fulfill requirements of warranty.

2. Identify related maintenance and reset procedure.

An example of a related item might be replacement of a water pump. Some water pumps are driven by the timing belt, and it may be wise to recommend replacing it at the same time as the timing belt is replaced. This would save the customer money as well as a potential unhappy customer, should a failure occur soon after the work was done. Another example might be the replacement of the fuel filter when an in-tank electric fuel pump is being replaced. Because fuel pumps often shed metal and field coil winding when they are on their way out, the fuel filter can become restricted and should be replaced at the same time. To recommend and sell related items benefits the customer; it also benefits the repair shop by providing an additional sale.

3. Locate and interpret maintenance schedule information.

Maintenance schedules are printed in the vehicle owner's manuals as well as in manufacturer's books and information systems. Most information systems contain a vehicle's maintenance schedules as well. Interpreting the schedule requires knowing the mileage and type of use the vehicle experiences, in addition to knowing what procedures are required to maintain the systems. If the cooling system, for example, must be serviced at 60,000 miles, what does that service involve? The manufacture may spell it out or leave it to standard procedures, in which case servicing the cooling system would include replacing the coolant and verifying its protection, inspecting the condition of all hoses and the radiator, and verifying the correct operation of the thermostat. Some vehicles may require other checkpoints based on the vehicle's history. Tracking history and reviewing it upon service visits help avoid mistakes in service suggestions.

6. *Warranty, Service Contracts, Service Bulletins, and Campaigns/Recalls (2 Questions)*

1. Demonstrate knowledge of warranty policies and procedures/parameters.

A service consultant should be very diligent about taking each customer very seriously and giving him/her complete attention. Many customers will claim to have some type of warranty or service policy that should cover the repair of their vehicle. In order to not make promises that will backfire, it is imperative that the service consultant be very knowledgeable about the various documents that may be used. It is advisable to read the details of each policy very carefully. Doing this will protect the service advisor and the repair shop from making a costly mistake. Once the details are understood by the service advisor, he/she should discuss the findings with the customer.

It is very crucial to thoroughly explain the procedures and parameters of the warranty to each customer. The customer needs to know exactly what they are responsible for paying and what the warranty will be paying for.

2. Locate and use reference information for warranties, service contracts, service bulletins, and campaigns/recalls.

Some service consultants must frequently work with service contracts. These are programs provided by various suppliers that extend the standard warranty. The process behind most of these programs is similar to working with an insurance company. The shop or the vehicle owner initiates a claim for needed repairs, an estimate is provided, the company providing the contract approves the part of the repair order that is covered, and the shop performs the work.

Warranty work and manufacturer campaigns/recalls are different. Warranty work may be paid for by the manufacturer, in the case of a new car warranty, or may be the responsibility of the shop, on a repair performed at its facility. Each business will have its own procedures for warranty.

When a manufacturer finds a persistent pattern of an engineering problem or failure of a specific component it may issue a campaign or a recall to repair or replace that component. Generally, these campaigns/recalls have an associated technical service bulletin (TSB) to help the shop performing the work understand the issue and the procedures associated with it. Since campaigns and recalls are generally paid for by the manufacturer, they are nearly always performed by the manufacturer's dealer.

There are several sources for information on campaigns and recalls. They include the manufacturer's information system, aftermarket information systems, trade magazines, the National Highway Traffic Safety Administration website, and customer letters sent by the manufacturer.

3. Demonstrate knowledge of service contract, technical service bulletin, and campaign/recall procedures to customers.

A professional service consultant needs to have a good understanding of technical documents that are involved in auto repair. These documents include service contracts, TSBs, and campaign/recall alerts.

A *service contract* is a type of extended warranty that many customers purchase when they buy their vehicle. These contracts are basically an insurance policy that pays for the repairs that are listed in the policy. The service consultant needs to be very careful when performing service on a vehicle that is covered by a service contract. It is almost always necessary to call the issuing company to get authorization to complete the repair. In addition, most of these policies carry a deductible that the customer must pay. It is important to clarify this fact with the customer prior to performing the repair.

A *technical service bulletin* is a document produced by the manufacturer to assist technicians in fixing pattern failures. It is important to clarify that these TSBs are not meant to be a "free fix" for the customer. They are just used to speed up the repair process by sharing learned knowledge about pattern failures with working technicians. If a vehicle is still covered by manufacturer's warranty, then the procedures followed in compliance with a TSB may be covered with no charge.

A *campaign/recall* is a policy/procedure to invite customers to bring their vehicle back in for a free modification. The key with these policies is that the manufacturer seeks to fix these problems as soon as possible due to safety-related concerns. Campaigns/recalls are always performed by a dealer for each manufacturer.

4. Verify the applicability of warranty, service contract, technical service bulletin, and campaign/recalls.

Using warranties, service contracts, TSBs, and recalls requires the service consultant to check for the process to be VIN-specific. It is important to make sure that the bulletin or campaign applies to the vehicle being worked on. It is also important to make sure that warranties and service contracts are in force for the vehicle before performing repair using them. This may require a call to the manufacturer or the contract provider. A wise service consultant keeps good documentation of every phase of the repair process in order to have an accurate account of what happened in each situation. This serves the consultant well if a customer has detailed questions about the repair process.

7. Vehicle Identification (2 Questions)

1. Locate and utilize vehicle identification number (VIN).

By the beginning of the 1980s, vehicles had become very diverse; even within the same body line there could be many variations. It had also become apparent that with vehicles sold in the United States coming from so many different origins, there was a greater risk that two completely different vehicles might end up with the same serial number. It was at this time that the 17-digit *vehicle identification number* (VIN) system we use today was adopted. The VIN became a code to identify the make, model, body, origin, year of production, engine size, and production sequence number. Using this code ensured that each vehicle would have a unique identification sequence. There are locations within the VIN that are specifically used to identify certain information. Others are left to the manufacturer to use as they wish. In analyzing some of the required VIN digits, the first position is used to denote the origin and manufacturer of the vehicle. Letters and numbers have been used. The eighth position is the engine for all domestic manufacturers, and the tenth digit is the year of production for the vehicle. The last six or seven digits are the serial number. It is very important that the VIN is recorded correctly and kept on file.

2. Locate the production date.

A service consultant should familiarize himself/herself with the location of the production date on the vehicle lines that visit the shop. Most manufacturers have a data plate or sticker inside the driver side doorjamb to give information about the vehicle. Some vehicles have the production date on a plate under the hood, usually on the fire wall. The various locations make it unlikely that the exam will include a question asking where a production date is. Correct part ordering requires this often forgotten information.

3. Locate and utilize component identification data.

Many vehicle components have tags or markings that identify them when it comes time for their replacement. They may be part numbers or engineering numbers cast into a part or a sticker on the component. They are often invaluable when ordering parts. A simple battery sticker can provide much information as to spec, size, group, and age.

A service consultant should have a good understanding of how to explain the repair process as well as show the customer where the parts in question are located. This knowledge assists the service consultant in explaining the labor charges to customers.

4. Identify body styles.

There are a number of ways to identify a vehicle's body style. For models with a limited number of variations available, the service consultant can simply identify it as a 2-door, 4-door sedan, station wagon, etc. Some manufacturers have more than 20 variations in a single body design. This is where it becomes important to use the VIN and the badges on the body to identify the vehicle. Modern software makes suggestions and provides drop-down displays to properly enter a vehicle.

5. Locate paint and trim codes.

As with the production date, the paint and trim codes are often found on the doorjamb or under the hood, on a plate. Some may even be located on the glove box door or the spare tire cover. The service consultant will need to be able to locate this information. Work with technicians to insist that they perform the necessary steps to preserve, not destroy, the above identifications. They become a critical part of the repair process.

C. Shop Operations (3 Questions)

Maintaining smooth shop operations is a task that a professional service consultant must achieve. It is very important to make sure that there is enough work to keep the service technicians busy. However, scheduling more work than can be completed in a day is not wise, because the customers will be unhappy when their vehicle does not get done when it was promised. The sections that follow will assist the service consultant in understanding the various skills that must be learned in order to run an efficient automotive repair facility.

1. Manage work flow.

The service consultant is in the unique position of knowing the customer's needs and the technician's work load. This puts him/her in the perfect position to manage the work flow in the shop. By keeping track of the amount of time necessary to complete a customer's vehicle and parts availability, the service consultant can determine the best times to promise completion to the customer. Once these expectations are set, it is important to stay on top of the work flow to be sure that technicians receive parts as expected and are able to complete the work on time. This is an area where a little communication can make the service consultant's job much easier. Shop efficiency relates directly to a shop's profitability or inability to support shop expenses.

2. Demonstrate knowledge of sublet procedures.

Many shops sublet operations that are not performed in their facilities. Common examples are driveline repair/balancing, transmission overhaul, body and paint work, or radiator repair. Sample questions will be provided in the practice tests. Knowledge builds trust through effective repairs.

3. Maintain customer appointment log.

Most of the repair shops in this country schedule appointments for their customers. When a customer calls for an appointment, the service consultant must determine what the needs are, about how much time it will take to repair the vehicle, and the availability of a technician(s) with adequate skills to complete the repair. An appointment log offers an organized method for scheduling, and also reminds the consultant of incoming work.

4. Address repeat repairs/comebacks.

Having to repair the same problem more than once can lower shop productivity and morale. When these problems present themselves, the service consultant is often called upon to get to the bottom of the problem and come up with an appropriate response. Is the technician missing the root problem for some reason? Is a part failing repeatedly? Does the customer keep bringing the vehicle back with a problem that appears to be resolved? Here is where the service consultant's question-asking expertise comes in. By interviewing the customer and the technician, he can determine the best course of action and explain it to all involved. These types of problems usually arise because a key piece of information is missing, such as a symptom that the customer left out or an unread TSB that outlines a procedure to alleviate the repeat problem.

The service consultant must work with the technician and the customer to find the problem's resolution. As with all things, doing a little research the first time the problem presents itself can save both the customer and the shop's resources later. Some modern-day problems do require a second or follow-up visit, so educating the customer on intermediate problems may result in a second chargeable invoice being sold with an incident. Always treat a comeback customer with the same courtesy that was provided when he or she first visited the shop.

SECTION

5 Sample Preparation Exams

INTRODUCTION

Included in this section are a series of six individual preparation exams that you can use to help determine your overall readiness to successfully pass the Service Consultant (C1) ASE certification exam. Located in Section 7 of this book you will find blank answer sheet forms you can use to designate your answers to each of the preparation exams. Using these blank forms will allow you to attempt each of the six individual exams multiple times without risk of viewing your prior responses.

Upon completion of each preparation exam, you can determine your exam score using the answer keys and explanations located in Section 6 of this book. Included in the explanation for each question is the specific task area being assessed by that individual question. This additional reference information may prove useful if you need to refer back to the task list located in Section 4 for additional support.

PREPARATION EXAM 1

1. Service Consultant A speaks clearly when having a conversation with a customer on the phone. Service Consultant B says that treating customers with dignity and respect on the phone is a positive business trait. Who is correct?

   A. A only
   B. B only
   C. Both A and B
   D. Neither A nor B

2. Service Consultant A asks the customer to speak very slowly to allow every comment to be written on the repair order. Service Consultant B asks open-ended questions when attempting to identify the customer concern to be written on the repair order. Who is correct?

   A. A only
   B. B only
   C. Both A and B
   D. Neither A nor B

3. Service consultants should use all of the following when greeting a new customer EXCEPT:

   A. Good eye contact
   B. Cordial handshake
   C. Customer's first name
   D. Genuine smile

4. Which of the following pieces of data can be determined from viewing the service history of a vehicle?

   A. Vehicle production date
   B. Service repair procedures performed at this location
   C. Location of the selling dealer
   D. Open recalls for the vehicle

5. A customer has an electrical problem with a vehicle. Service Consultant A promises a completion time to the customer during the write-up process. Service Consultant B gives the customer regular updates on the status of his/her vehicle throughout the repair visit. Who is correct?

   A. A only
   B. B only
   C. Both A and B
   D. Neither A nor B

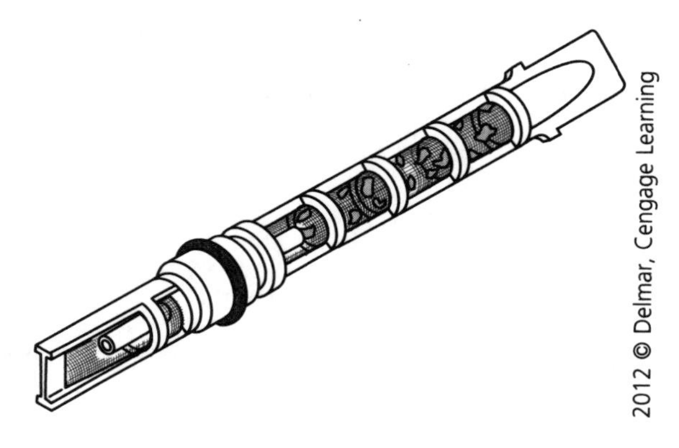

6. Service Consultant A says that the component shown in the figure above needs to be serviced and that the A/C system from which it came needs to be flushed. Service Consultant B says that a failed A/C compressor could have caused the debris on the screen. Who is correct?

   A. A only
   B. B only
   C. Both A and B
   D. Neither A nor B

7. Service Consultant A says that when writing up a comeback/warranty ticket, it is necessary to review previous repair orders with the customer. Service Consultant B says that when writing up a comeback/warranty ticket, it is necessary to ask the customer to restate the symptoms he is experiencing. Who is right?

   A. A only
   B. B only
   C. Both A and B
   D. Neither A nor B

8. All of the following are positive results from a customer follow-up call EXCEPT:

    A. Constructive criticism is gained from customer comments.
    B. Shop income can be increased due to new appointments.
    C. The service consultant can confront the customer who has negative feedback.
    D. The customer will appreciate the follow-up call.

9. A vehicle had a new water pump installed just 30 days ago and is back in the repair shop with the water pump bearing making a growling noise. Service Consultant A recognizes that this condition should be covered under the parts warranty of the previously installed water pump. Service Consultant B says that the customer should have to pay for this repair again because engine parts do not carry any warranty coverage. Who is correct?

    A. A only
    B. B only
    C. Both A and B
    D. Neither A nor B

10. Service Consultant A says that a hard-top convertible will have limited trunk space when the top is down due to the vehicle design. Service Consultant B says that a soft-top convertible will typically have a more rigid body and frame due to the absence of the structural top. Who is correct?

    A. A only
    B. B only
    C. Both A and B
    D. Neither A nor B

11. Service Consultant A says that major engine components are typically covered by extended warranties. Service Consultant B says that brake pads are typically covered by extended warranties. Who is correct?

    A. A only
    B. B only
    C. Both A and B
    D. Neither A nor B

12. Service Consultant A says that giving a ballpark estimate during the write-up process is a good idea. Service Consultant B says that repair estimates should only be given after the technician has diagnosed the vehicle and checked the cost of the parts needed. Who is correct?

    A. A only
    B. B only
    C. Both A and B
    D. Neither A nor B

13. Service Consultant A recommends that the vehicle tires be rotated at every other oil change. Service Consultant B recommends that the vehicle tire pressure be checked at every oil change. Who is correct?

    A. A only
    B. B only
    C. Both A and B
    D. Neither A nor B

14. Service Consultant A says that some engines use a timing chain to connect the crankshaft to the camshaft. Service Consultant B says that the crankshaft and camshaft turn at the same speed when the engine is running. Who is correct?
    A. A only
    B. B only
    C. Both A and B
    D. Neither A nor B

15. Service Consultant A wears jeans and a worn shirt to work. Service Consultant B wears clean "business casual" clothes and always tucks in his/her shirt while at work. Who is correct?
    A. A only
    B. B only
    C. Both A and B
    D. Neither A nor B

16. A vehicle with 59,985 miles is in a service facility for a safety inspection prior to a pending trip. Service Consultant A says it is advisable to recommend that the timing chain be replaced due to the mileage. Service Consultant B says it is wise to recommend a 60,000-mile service be performed before the trip rather than waiting until after. Who is correct?
    A. A only
    B. B only
    C. Both A and B
    D. Neither A nor B

17. All of the following are positive characteristics of a well-run service facility EXCEPT:
    A. Clean and comfortable waiting area
    B. Employees who are knowledgeable and efficient
    C. Location on a busy street
    D. Neat and organized write-up area

18. Service Consultant A says that obtaining accurate contact information is usually not necessary. Service Consultant B says that most customers expect the service facility to provide some form of transportation if the service takes longer than they can readily wait. Who is correct?
    A. A only
    B. B only
    C. Both A and B
    D. Neither A nor B

19. Which of the following would be the highest priority service on a late-model vehicle?
    A. Hub cap replacement
    B. Steering gear replacement
    C. Cabin filter replacement
    D. Radio replacement

20. Which engine component regulates engine temperature?
    A. Heater core
    B. Thermostat
    C. Water pump
    D. By-pass hose

21. Which fuel system component sprays fuel into the intake manifold when it receives a signal from the power train control module (PCM)?
    A. Fuel pump
    B. Fuel pressure regulator
    C. Fuel injector
    D. Fuel sending unit

22. Shop quality control is a very important aspect of a successful service facility. Service Consultant A encourages her technicians to keep their work areas as clean and orderly as possible. Service Consultant B continually scans the shop area to check on the status of the vehicles. Who is practicing the correct activities?
    A. A only
    B. B only
    C. Both A and B
    D. Neither A nor B

23. Each of the following procedures would be performed during a typical 60,000-mile service EXCEPT:
    A. Replace the oil pump.
    B. Replace the spark plugs.
    C. Replace the fuel filter.
    D. Inspect and adjust the brake system.

24. All of the following components are part of a typical starting system EXCEPT:
    A. Starter solenoid
    B. Park/neutral switch
    C. Ignition switch
    D. Voltage regulator

25. Which drive train component connects the transmission to the drive axle on a rear-wheel drive vehicle?
    A. Axle half shaft
    B. Driveshaft
    C. Universal joint
    D. Carrier bearing

26. Service Consultant A says that items such as belts and hoses are considered wear items and are not typically covered by extended warranties. Service Consultant B says that major transmission components are considered wear items and are not typically covered by extended warranties. Who is correct?
    A. A only
    B. B only
    C. Both A and B
    D. Neither A nor B

27. Service Consultant A says that the 10th digit of the VIN represents the year model. Service Consultant B says that the 8th digit of the VIN is the engine code. Who is correct?
    A. A only
    B. B only
    C. Both A and B
    D. Neither A nor B

28. Which of the following is the most common method of communicating the customer request to the technician?

    A. Sending a text message to the technician with the customer request.
    B. Announcing the information over the intercom.
    C. Having the vehicle porter relay to the technician the details revealed from a follow-up phone call.
    D. Writing clear and complete customer concern descriptions on the repair order.

29. Which automatic transmission component acts as a fluid-coupling device between the engine and transmission?

    A. Transmission oil pump
    B. Torque converter
    C. Clutch pack
    D. Valve body

30. Service Consultant A says that some repair shops sublet radiator repairs to a specialty shop. Service Consultant B says that some repair shops sublet coolant maintenance repairs to a specialty shop. Who is correct?

    A. A only
    B. B only
    C. Both A and B
    D. Neither A nor B

31. Service consultants must strive to keep an accurate customer appointment log. Service Consultant A says that it is advisable to schedule more work than the shop can complete in order to make sure the day is filled with plenty of jobs. Service Consultant B says that it is advisable to schedule only 50 percent of the shop's capacity for a day in order to make sure that all vehicles get completed on time. Who is correct?

    A. A only
    B. B only
    C. Both A and B
    D. Neither A nor B

32. Service Consultant A says that some shops require the technician to clock in on the repair order when he begins and when he finishes a repair. Service Consultant B says that some shops monitor the hours that each technician produces each week in order to track technician efficiency. Who is correct?

    A. A only
    B. B only
    C. Both A and B
    D. Neither A nor B

33. Service Consultant A says that it is wise to post the certifications and credentials of the service technicians in the write-up area. Service Consultant B says that most customers are more worried about low cost service than about the professionalism of the service technicians. Who is correct?

    A. A only
    B. B only
    C. Both A and B
    D. Neither A nor B

34. Service Consultant A always provides vague estimates to his customers so he can quickly get the technician started on the repair. Service Consultant B provides thorough and accurate estimates to her customers so there will not be any surprises when it is time to pay the bill. Who is correct?

A. A only
B. B only
C. Both A and B
D. Neither A nor B

35. Supplemental restraint system (SRS) is another name for:

A. Seat belts
B. Airbags
C. Upper hydraulic motor mount
D. Child safety seat anchors

36. Service Consultant A says that an example of a feature of an oil change is the brand of oil used. Service Consultant B says that an example of a benefit of an oil change is longer engine life. Who is right?

A. A only
B. B only
C. Both A and B
D. Neither A nor B

37. Service Consultant A says that adding a description of the work performed adds value to the repair. Service Consultant B says that the VIN number may be used to find part applications. Who is right?

A. A only
B. B only
C. Both A and B
D. Neither A nor B

38. Service Consultant A says that the oil change reminder light should be reset by the technician when the engine oil is changed. Service Consultant B says that the oil life percentage reminder system should be reset when the oil is changed. Who is correct?

A. A only
B. B only
C. Both A and B
D. Neither A nor B

39. All of the following are elements of an oil change service EXCEPT:

A. Test the pH level of the coolant.
B. Drain the oil and refill with the correct type and amount of new oil.
C. Remove the oil filter and replace with the correct new oil filter.
D. Check all engine fluid levels and notify the owner of the systems that were low.

40. Which hydraulic brake component converts linear movement into hydraulic pressure that is sent to the wheels to stop the vehicle?

    A. Metering valve
    B. Proportioning valve
    C. Wheel cylinder
    D. Master cylinder

41. Service Consultant A says that camber is the alignment angle that measures the inward or outward tilt of the tire. Service Consultant B says that caster is the alignment angle that measures the forward or rearward tilt of the steering axis. Who is correct?

    A. A only
    B. B only
    C. Both A and B
    D. Neither A nor B

42. Which of the following items would be the LEAST LIKELY component to be replaced during a clutch replacement?

    A. Throw out bearing
    B. Pressure plate
    C. Input gear
    D. Clutch disc

43. Which of the ignition system components creates the high-voltage spike that creates the energy to fire the spark plug?

    A. Ignition module
    B. Coil pack
    C. Crank sensor
    D. Plug wire

44. Service Consultant A says that the electric fuel pump is located inside the fuel tank. Service Consultant B says that the fuel filter is located under the intake manifold plenum. Who is correct?

    A. A only
    B. B only
    C. Both A and B
    D. Neither A nor B

45. Service Consultant A carefully documents the services that are performed on each repair order prior to notifying the customer that the vehicle is done. Service Consultant B documents the recommended services on the repair order in order to communicate this information to the customer. Who is correct?

    A. A only
    B. B only
    C. Both A and B
    D. Neither A nor B

46. All of the following examples would be considered a benefit of recommending additional services to a current customer who has left his/her car at your shop EXCEPT:

    A. The shop can stay open late to finish the repair.
    B. The shop is more profitable from the increased sales of parts.
    C. The shop is more profitable from the increased sales of labor.
    D. The customer will begin to trust that the shop is looking out for his/her well-being and safety.

47. A customer is objecting to having additional needed maintenance service performed. Service Consultant A explains that the vehicle will likely break down very soon and strand the customer in a dangerous location. Service Consultant B explains the value of having a well-maintained vehicle. Who is correct?

    A. A only
    B. B only
    C. Both A and B
    D. Neither A nor B

48. A customer is at the repair shop describing the problems with his/her vehicle. Service Consultant A asks detailed questions of the customer to determine what the main concern is. Service Consultant B asks questions about the time, temperature, and the frequency with which the problem is occurring. Who is correct?

    A. A only
    B. B only
    C. Both A and B
    D. Neither A nor B

49. Which of the following would be the most critical safety service to be performed on a late-model vehicle?

    A. Oil change
    B. Door panel replacement
    C. Brake pad replacement
    D. Air conditioner service

50. All of the following would be benefits of having service repair work done at the same quality repair shop every time EXCEPT:

    A. The oil changes are cheaper.
    B. A service history is developed for the vehicle.
    C. A trustful relationship is developed between the owner and the repair shop's employees.
    D. The technicians become familiar with the vehicle.

PREPARATION EXAM 2

1. Service Consultant A says that a ringing phone always takes precedent over a customer standing in front of him. Service Consultant B says that if a customer walks in while you are on the phone, you should quickly take a message and deal with the customer in front of you. Who is correct?
   - A. A only
   - B. B only
   - C. Both A and B
   - D. Neither A nor B

2. All of the following vehicle information items should be collected when preparing a vehicle repair order EXCEPT:
   - A. VIN
   - B. Color
   - C. Transmission model
   - D. Vehicle model

3. Which of the following items is NOT listed as a digit on the VIN?
   - A. Country of origin
   - B. Body style
   - C. Vehicle year
   - D. Vehicle build month

4. A vehicle has been dropped off at the repair shop with instructions from the owner to diagnose a brake noise and then call with an estimate and completion time. Service Consultant A checks the appointment log and the status of other jobs with the technician before predicting a completion time. Service Consultant B verifies the price and availability of the parts before predicting a completion time. Who is correct?
   - A. A only
   - B. B only
   - C. Both A and B
   - D. Neither A nor B

5. Which engine component will benefit the most from regular oil and filter changes?
   - A. Thermostat
   - B. Cylinder head
   - C. Crankshaft
   - D. Piston pin

6. All of the following are positive reasons a service consultant could use to finalize a service sale EXCEPT:
   - A. The vehicle will be more reliable if the correct service and maintenance are performed on time.
   - B. The owner will be more likely to trade the vehicle because he is dissatisfied with it.
   - C. The vehicle will hold its value better if good service practices are followed.
   - D. The owner will be safer driving a well-maintained vehicle.

7. Service Consultant A says that some front-wheel drive vehicles have a full steel frame that is similar to rear-wheel drive vehicles. Service Consultant B says that some front-wheel drive vehicles use an electrically assisted power steering system. Who is correct?

   A. A only
   B. B only
   C. Both A and B
   D. Neither A nor B

8. Service Consultant A does not check for the availability of the repair parts before notifying the customer. Service Consultant B does not call the customer until after checking the availability and price of the repair parts. Who is correct?

   A. A only
   B. B only
   C. Both A and B
   D. Neither A nor B

9. Service Consultant A says that the power windows operate by using bi-directional solenoids to move the window up and down. Service Consultant B says that the power seats use bi-directional permanent magnet (PM) motors to move the seat to the desired location. Who is correct?

   A. A only
   B. B only
   C. Both A and B
   D. Neither A nor B

10. All of the following questions should be asked on a customer follow-up call EXCEPT:

   A. Were you pleased with the way you were greeted at the repair shop?
   B. Were you pleased with the neatness of the waiting area at the repair shop?
   C. Were you pleased with the brand of coffee that was served in the waiting room?
   D. Were you pleased with the quality of the repair that was performed by the facility?

11. Service Consultant A says that the vehicle battery cable should never be disconnected while the engine is running. Service Consultant B says the vehicle battery can run down if the vehicle is stored for long periods of time. Who is correct?

   A. A only
   B. B only
   C. Both A and B
   D. Neither A nor B

12. All of the following items should be added to the estimate when calculating the total cost for a repair EXCEPT:

   A. Labor amount paid to the technician
   B. Parts total
   C. Labor total
   D. Sublet repair costs

13. A customer is requesting information about the warranty of the work that is performed by a repair shop. Service Consultant A says that some remanufactured engines carry a warranty for as long as 3 months or 3,000 miles. Service Consultant B says that some batteries carry a three-year free replacement warranty. Who is correct?

    A. A only
    B. B only
    C. Both A and B
    D. Neither A nor B

14. Service Consultant A is offended if a customer asks detailed questions about why a particular service is being recommended. Service Consultant B provides clear and understandable answers to detailed questions. Who is correct?

    A. A only
    B. B only
    C. Both A and B
    D. Neither A nor B

15. What is the LEAST LIKELY step that would be performed during an automatic transmission fluid and filter service?

    A. Install new friction discs into the clutch packs.
    B. Remove the transmission oil pan and drain the old fluid into a suitable container.
    C. Replace or clean the old transmission filter.
    D. Install the transmission pan with a new gasket.

16. Service Consultant A says that it is not necessary to know what has been done to the vehicle in the past. Service Consultant B says that some shops have computerized software that stores the service data on each vehicle that has been in the repair facility. Who is correct?

    A. A only
    B. B only
    C. Both A and B
    D. Neither A nor B

17. All of the following items require a reset procedure by a repair shop after services are performed EXCEPT:

    A. Oil change reminder light
    B. Oil life percentage reminder
    C. Check engine light
    D. Low coolant level indicator

18. Which of the following repair procedures would be the LEAST LIKELY to be considered a high priority repair?

    A. Replacement of a worn tie rod end
    B. Replacement of the rear wiper blade
    C. Replacement of a tire with the steel showing
    D. Front brake pads replaced

19. Service Consultant A says that a manufacturer recall/campaign is a technical document created to help technicians to repair pattern failures more quickly. Service Consultant B says that a technical service bulletin (TSB) is a program that manufacturers create to invite customers to bring their vehicles back in for a free repair in order to correct a safety fault in the vehicle. Who is correct?
    A. A only
    B. B only
    C. Both A and B
    D. Neither A nor B
20. A first-time customer walks into a repair shop and asks the service consultants at the desk to explain the benefits and features of having repair work done at that shop. Service Consultant A says that one benefit of the shop is that all of the technicians are ASE certified in the areas in which they perform repair work. Service Consultant B says that the employees at the shop are more honest than at other shops in town. Who is correct?
    A. A only
    B. B only
    C. Both A and B
    D. Neither A nor B
21. All of the following items should be entered into the customer appointment log EXCEPT:
    A. Tire size
    B. Vehicle year, make, and model
    C. Estimated time of repair
    D. Customer name
22. What is the most likely part that would be serviced when the technician states that the engine needs a tune up?
    A. Piston rings
    B. Valve cover gaskets
    C. Fuel filter
    D. Oxygen sensor
23. Service Consultant A verifies whether a customer is a first-time customer when scheduling an appointment. Service Consultant B verifies whether a customer is a potential warranty customer when scheduling an appointment. Who is correct?
    A. A only
    B. B only
    C. Both A and B
    D. Neither A nor B
24. All of the following are appropriate greeting practices for a service consultant EXCEPT:
    A. Smile
    B. Cordial attitude
    C. Handshake
    D. Fist bump
25. Service Consultant A feels that she is not receiving the support from the manager that she needs, so she is usually negative with the technicians and customers. Service Consultant B has several items that concern him about the support he is receiving from the manager, so he schedules time for a face-to-face meeting with the manager. Who is correct?
    A. A only
    B. B only
    C. Both A and B
    D. Neither A nor B

26. A vehicle is in the repair shop with a check engine light that stays illuminated. The technician has checked the computer system for a problem and diagnosed the problem to be a faulty oxygen sensor. Service Consultant A says that oxygen sensors measure the amount of oxygen in the air as it passes through the intake manifold. Service Consultant B says that the oxygen sensor can deteriorate over time and requires periodic replacement. Who is correct?
    A. A only
    B. B only
    C. Both A and B
    D. Neither A nor B

27. From which source does the vehicle electrical system receive its energy?
    A. Starter
    B. Engine control module (ECM)
    C. Alternator
    D. Body control module (BCM)

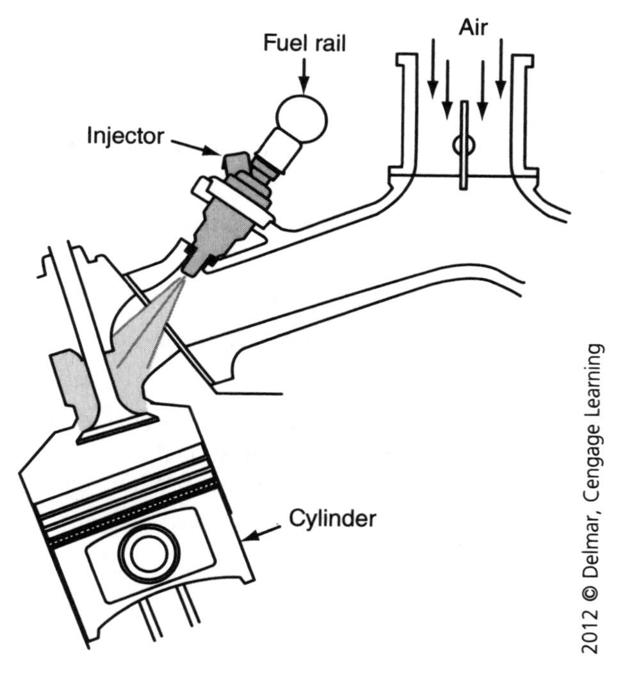

28. All of the following statements are correct about the figure above EXCEPT:
    A. The electric fuel pump supplies pressurized fuel to the fuel injector.
    B. The throttle plate opens and closes as the accelerator pedal is depressed.
    C. The injector supplies fuel for all of the cylinders on the engine.
    D. The injector sprays fuel, which mixes with the air before entering the combustion chamber.

29. A customer visits a repair facility with a transmission problem. The transmission is not shifting correctly and the repair will likely take several hours. Service Consultant A offers to let the customer use the phone to call someone to pick him/her up. Service Consultant B directs the customer to the waiting room. Who is correct?
    A. A only
    B. B only
    C. Both A and B
    D. Neither A nor B

30. A customer arrives to pick her vehicle after extensive repairs are made. Service Consultant A directs the customer to the cashier to pay her bill and has the vehicle porter pull the vehicle around to the front of the building. Service Consultant B carefully explains all of the repairs and charges to the customer before directing her to the cashier to pay the bill. Who is correct?

    A. A only
    B. B only
    C. Both A and B
    D. Neither A nor B

31. Service Consultant A says that the engine flywheel drives the clutch disc when the clutch pedal is pressed. Service Consultant B says that the pressure plate is the part of the system that compresses the clutch disc into the flywheel when the clutch pedal is not pressed. Who is correct?

    A. A only
    B. B only
    C. Both A and B
    D. Neither A nor B

32. Each of the following is a benefit of recommending additional services to a current customer who has left his/her car at your shop EXCEPT:

    A. The customer will need a second opinion.
    B. The shop is more profitable from the increased sales of parts.
    C. The shop is more profitable from the increased sales of labor.
    D. The customer will begin to trust that the shop is looking out for his or her well-being and safety.

33. Service Consultant A says that all the automatic transmission fluid should be changed once each year or every 12,000 miles. Service Consultant B says that all transmission fluid is designed to last for the life of the vehicle if the vehicle is not used to pull a heavy load. Who is correct?

    A. A only
    B. B only
    C. Both A and B
    D. Neither A nor B

34. Which of the following brake system components converts linear movement into hydraulic pressure that is sent to the wheel brakes?

    A. Wheel cylinder
    B. Vacuum booster
    C. Caliper
    D. Master cylinder

35. What are the most likely steps that would be followed by a service consultant to verify the accuracy of a repair order prior to calling the customer for repair authorization?

    A. Add the labor and the parts on the back of the repair order.
    B. Input all the repair-related data into the computer and verify the availability of the parts.
    C. Add the cost of the parts, the cost of the labor, and the cost of sublet repairs.
    D. Input the parts cost, labor cost, tax cost, and the sublet repair cost into a calculator.

36. Service Consultant A typically recommends that the alignment be checked prior to having new tires installed to assure longer lasting tire life. Service Consultant B only recommends the lowest price tires available in order to save the customer some expense. Who is correct?
    A. A only
    B. B only
    C. Both A and B
    D. Neither A nor B

37. Which of the following service consultant choices for attire would be LEAST LIKELY to present a professional image?
    A. Green shirt and blue pants
    B. Red shirt and blue jeans
    C. White shirt and tan pants
    D. Clean and neatly groomed hair

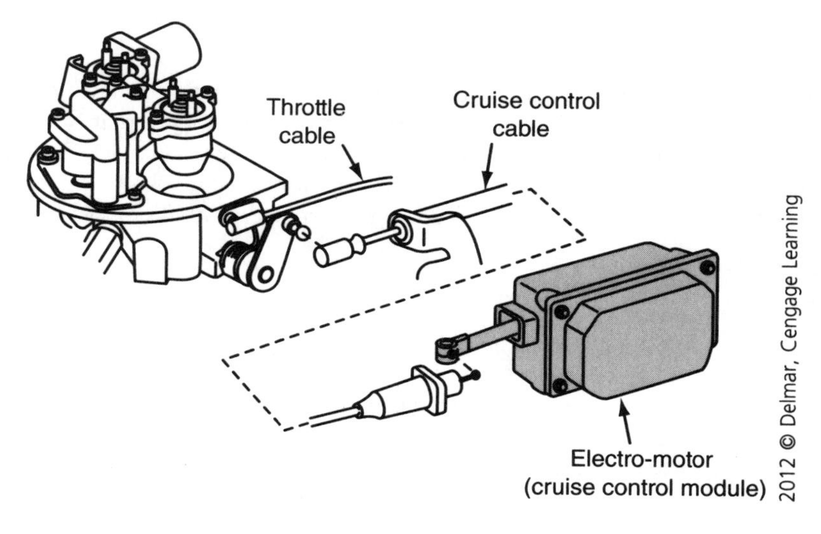

38. All of the following statements concerning the figure above are correct EXCEPT:
    A. The cruise control system must disengage when the brakes are applied.
    B. The cruise control system must disengage when the clutch is applied.
    C. The cruise control module is mounted to the cruise electromotor.
    D. The cruise motor must have a vacuum hose connected in order to work properly.

39. The refrigerant changes from a high-pressure vapor to a high-pressure liquid as it passes through which A/C component?
    A. Evaporator
    B. Compressor
    C. Receiver/drier
    D. Condenser

40. All of the following procedures would be performed during a typical 60,000-mile service EXCEPT:
    A. Replace the rear main seal.
    B. Replace the spark plugs.
    C. Replace the fuel filter.
    D. Inspect and adjust the brake system.

41. Service Consultant A takes very little time explaining the estimate due to the hurried pace of the repair shop. Service Consultant B calculates an estimate and then adds a small percentage to cover incidental costs that might arise. Who is correct?

A. A only
B. B only
C. Both A and B
D. Neither A nor B

42. Service Consultant A says that it is wise to group several maintenance items into a package to increase the sense of good value to the customer. Service Consultant B says that a maintenance menu board should be posted in a convenient location near the service write-up area. Who is correct?

A. A only
B. B only
C. Both A and B
D. Neither A nor B

43. A vehicle is in the repair shop for an inspection. The technician recommends having the tires balanced for a vibration and the clock spring replaced due to an airbag light that is illuminated. The customer asks the service consultant to prioritize the two services in order of importance. Service Consultant A recommends the tire rotation since it is less expensive. Service Consultant B says that both items are of equal importance. Who is correct?

A. A only
B. B only
C. Both A and B
D. Neither A nor B

44. A vehicle is in the repair shop for a thermostat replacement. Service Consultant A recommends that the cooling system be totally flushed and refilled with new coolant while the thermostat is being replaced. Service Consultant B recommends that the cooling system be leak-tested with a pressure tester while the thermostat is being replaced. Who is correct?

A. A only
B. B only
C. Both A and B
D. Neither A nor B

45. All of the following services may be considered a sublet repair for some mechanical shops EXCEPT:

A. Radiator repair
B. Air filter replacement
C. Painting a body panel
D. Transmission overhaul

46. Service Consultant A says that you should directly quote the customer's comments on the repair order. Service Consultant B says that you should ask probing questions about when, where, and how the problem occurs and paraphrase the customer's comments on the repair order. Who is correct?

A. A only
B. B only
C. Both A and B
D. Neither A nor B

47. Where is the most likely location of the vehicle production date?

 A. On the B-pillar on the driver's side
 B. In the trunk near the spare tire
 C. On the B-pillar on the passenger's side
 D. On the engine data tag

48. Service Consultant A says that it is critical to have an accurate estimate before calling the customer for approval. Service Consultant B says that a reasonable completion time is important to have before calling for customer approval. Who is correct?

 A. A only
 B. B only
 C. Both A and B
 D. Neither A nor B

49. All of the following would be normal activities in a well-run service facility EXCEPT:

 A. A customer enters the service write-up area and is quickly greeted by a service consultant.
 B. Technicians are working on customers' vehicles with several radios turned up loudly.
 C. Vehicle porters are putting seat covers, steering wheel covers, and floor mat covers into each car as it enters the service write-up area.
 D. Customers who are waiting for their vehicles have a clean and comfortable waiting area.

50. Service Consultant A strives to keep an accurate customer appointment log by always writing the appointments down. Service Consultant B only schedules appointments to 50 percent of the shop's capacity so that the shop is not overbooked. Who is correct?

 A. A only
 B. B only
 C. Both A and B
 D. Neither A nor B

PREPARATION EXAM 3

1. The telephone is ringing while a service consultant is working with a customer at the write-up desk. Service Consultant A does not answer the phone and lets the call go to voice mail. Service Consultant B answers the phone and engages in a 12-minute conversation. Who is correct?

 A. A only
 B. B only
 C. Both A and B
 D. Neither A nor B

2. All of these are components of the charging system EXCEPT:

   A. Starter
   B. Serpentine belt
   C. Voltage regulator
   D. Alternator

3. Service Consultant A says that providing a ballpark estimate is a useful tool for closing a sale. Service Consultant B says that asking for an appointment is a good way to close a sale. Who is correct?

   A. A only
   B. B only
   C. Both A and B
   D. Neither A nor B

4. Service Consultant A says that a sedan has four doors. Service Consultant B says that a coupe has two doors. Who is correct?

   A. A only
   B. B only
   C. Both A and B
   D. Neither A nor B

5. Service Consultant A says that customers will sometimes ask for alternative transportation when they make an appointment. Service Consultant B says that most customers do not expect the service facility to provide some form of transportation if the service takes longer than they can readily wait. Who is correct?

   A. A only
   B. B only
   C. Both A and B
   D. Neither A nor B

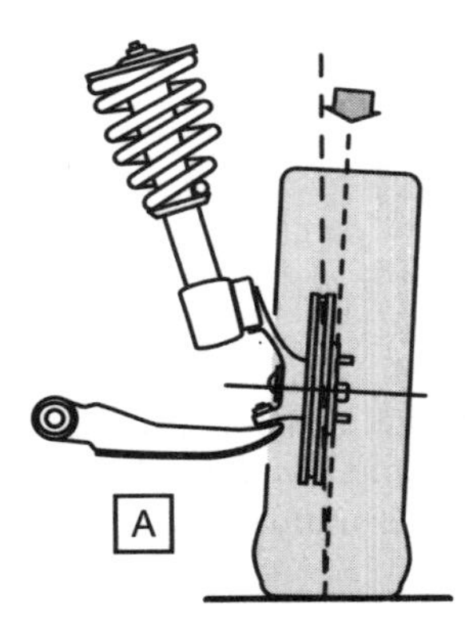

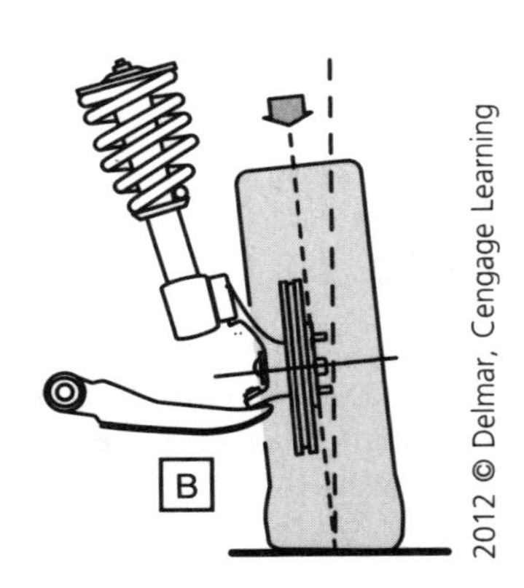

6. Which wheel alignment angle is indicated in the figure?

   A. Thrust angle
   B. Camber
   C. Caster
   D. Toe

7. Which of the following pieces of data can be determined from viewing the service history of a vehicle in the repair shop's computer system?

   A. A list of the service repair procedures performed at this location
   B. Vehicle production date
   C. The name of the salesperson who sold the car when it was new
   D. A list of service repair procedures at other repair shops

8. A customer receives a letter from the manufacturer for which of these actions?

   A. A technical service bulletin (TSB)
   B. A pattern failure
   C. End of vehicle warranty
   D. A manufacturer recall

9. Service Consultant A says that telling customers when their vehicle will be ready at the time they drop off the vehicle creates expectations. Service Consultant B says that accurate completion times can only be determined after vehicle inspection. Who is correct?

   A. A only
   B. B only
   C. Both A and B
   D. Neither A nor B

10. All of these systems use a filter EXCEPT:

    A. The automatic transmission
    B. The brake master cylinder
    C. The air conditioning (A/C) system
    D. The HVAC air handling system

11. What is the most likely benefit of performing a follow-up phone call after a repair visit?

    A. Seek referrals for new business from the customer.
    B. Offer discount coupons to the customer for them to give to friends and family.
    C. Offer discount repair services to the customer for returning to the shop for their next repair.
    D. Measure the satisfaction of the customer concerning the visit to the repair shop.

12. Service Consultant A always checks for parts availability before calling the customer with an estimate and completion time. Service Consultant B assumes that parts will be available and calls the customer in order to get the technician started on the repair as soon as possible. Who is correct?

    A. A only
    B. B only
    C. Both A and B
    D. Neither A nor B

13. Service Consultant A says that providing an estimate is required by law in some states. Service Consultant B says that explaining the details of the estimate helps to add value to the services the customer is buying from the shop. Who is correct?

    A. A only
    B. B only
    C. Both A and B
    D. Neither A nor B

14. The starter turns the engine by engaging with which of these components?

    A. Camshaft
    B. Flywheel ring gear
    C. Crankshaft
    D. Battery

15. Service Consultants need to be ready to answer questions from repair customers clearly. Service Consultant A provides clear and understandable answers to detailed questions that the customers have. Service Consultant B asks the customers if they have any questions before completing the write-up process. Who is correct?

    A. A only
    B. B only
    C. Both A and B
    D. Neither A nor B

16. Service Consultant A says that maintenance schedules are not relevant because it is impossible to estimate which driving style applies to the vehicle. Service Consultant B says that it is wise to follow the maintenance schedule that is closest to the driving style of each customer. Who is correct?

    A. A only
    B. B only
    C. Both A and B
    D. Neither A nor B

17. Repair shops work very hard to earn the trust of customers so that they will continue to patronize their location. Service Consultant A says that repeat customers receive more thorough service because all of the service records will be at one location. Service Consultant B says that repeat customers receive more thorough service because the repair technicians become familiar with the vehicle. Who is correct?

    A. A only
    B. B only
    C. Both A and B
    D. Neither A nor B

18. A customer recites a list of symptoms to the service consultant. What is the most likely next step that the service consultant would do?

    A. Write down exactly what the customer says.
    B. Use his/her experience to estimate repairs.
    C. Offer suggestions about what the problem might be.
    D. Ask open-ended questions to determine customer needs.

19. Providing accurate estimates is necessary for customers to make informed decisions about the repair of their vehicle. Service Consultant A carefully adds all of the charges together and then relays the information to the customer. Service Consultant B gives the customers a 25 percent discount in order to show the customer the value of the service being performed. Who is correct?

    A. A only
    B. B only
    C. Both A and B
    D. Neither A nor B

20. Interpreting the technician's diagnosis is an important skill for a service consultant. Service Consultant A does not try to thoroughly understand each diagnosis a technician makes; he just provides vague explanations to the customers with whom he communicates. Service Consultant B is fairly knowledgeable about most vehicle systems and strives to interpret each technician diagnosis and then relay the information in a way that the customer will understand it. Who is correct?

    A. A only
    B. B only
    C. Both A and B
    D. Neither A nor B

21. A customer brings a car to the repair shop with a complaint that the steering wheel vibrates while braking. Service Consultant A shows the customer all of the service coupon specials that are currently being offered by the shop. Service Consultant B tells the customer that the tires will need to be balanced to repair the problem. Who is correct?

    A. A only
    B. B only
    C. Both A and B
    D. Neither A nor B

22. All of the following information should be on the customer appointment log EXCEPT:

    A. Customer name
    B. Estimated time of repair
    C. Vehicle color
    D. Vehicle year, make, and model

23. The master cylinder is part of which system?

    A. Brake system
    B. Engine control system
    C. Automatic transmission system
    D. Power steering system

24. Which of the following repairs would be the highest priority for a customer to repair in terms of the vehicle's safety?

    A. Replace the rear wiper blade.
    B. Service the automatic transmission fluid and filter.
    C. Replace a tire with a cut in the sidewall.
    D. Service an inoperative A/C system.

25. A customer arrives to pick up his/her vehicle after extensive repairs are made. Service Consultant A calls the service porter to pull the vehicle around to the pickup area. Service Consultant B carefully explains all of the repairs and charges and asks the customer if there are any questions. Who is correct?

    A. A only
    B. B only
    C. Both A and B
    D. Neither A nor B

26. Service Consultant A says that a technician's efficiency and speed should be monitored to determine how much work to schedule for him on the appointment log. Service Consultant B says that the quality of the technician's work should be monitored and communicated to management. Who is correct?

    A. A only
    B. B only
    C. Both A and B
    D. Neither A nor B

27. All of the items below would be needed when collecting vehicle information for the vehicle repair order EXCEPT:

    A. Vehicle identification number (VIN)
    B. Mileage
    C. Tire size
    D. Vehicle make, model, and color

28. Which item connects to the connecting rod at the end opposite the crankshaft?

    A. The flywheel
    B. The rear main seal
    C. The piston
    D. The cylinder head

29. A technician performs an extensive inspection and recommends the following: replacement of a damaged driver-side seat belt, cooling system flush, replacement of brake pads that have 7/32 inch remaining, and an oil change that is 1,800 miles overdue. Which of these represents the best way to prioritize this list to the customer?

    A. Brake pads, oil change, seat belt, cooling system service
    B. Seat belt, oil change, brake pads, cooling system service
    C. Seat belt, brake pads, oil change, cooling system service
    D. Oil change, cooling system service, brake pads, seat belt

30. Service Consultant A says that the thermostat is a component of the engine cooling system. Service Consultant B says the heater core is a component of the engine cooling system. Who is correct?

    A. A only
    B. B only
    C. Both A and B
    D. Neither A nor B

31. Service Consultant A says that the torque converter is a component of the exhaust system. Service Consultant B says that catalytic converter is a component of the exhaust system. Who is correct?

A. A only
B. B only
C. Both A and B
D. Neither A nor B

32. All of these are components of an automatic transmission EXCEPT:

A. Planetary gear sets
B. Pressure plate
C. Torque converter
D. Valve body

33. Service Consultant A does not groom her hair while at work. Service Consultant B wears clean casual clothes but does not tuck in his shirt while at work. Who is correct?

A. A only
B. B only
C. Both A and B
D. Neither A nor B

34. An automatic transmission is being replaced. Service Consultant A says that the transmission cooler should be flushed prior to connecting to the transmission. Service Consultant B says that the engine oil must be changed to guarantee good transmission life. Who is correct?

A. A only
B. B only
C. Both A and B
D. Neither A nor B

35. A customer calls the repair shop to add a service request for a vehicle that is currently receiving service in the shop. Service Consultant A uses the intercom to relay the message to the technician. Service Consultant B sends a text message to the technician giving the customer's instructions. Who is correct?

A. A only
B. B only
C. Both A and B
D. Neither A nor B

36. Which of the following type of brake fluid has the lowest boiling point?

A. DOT 4
B. DOT 5
C. DOT 5.1
D. DOT 3

37. Which of the following is most likely to have the greatest impact on a customer's decision to do business with a repair shop?

    A. Extended business hours
    B. The service consultant's appearance
    C. The level of trust they feel
    D. Discount pricing

38. Which of these is a component of the supplemental restraint system (SRS) or airbag system?

    A. Throttle sensor
    B. Power train control module (PCM)
    C. Clock spring/spiral cable
    D. Wheel speed sensor

39. All of the following components are needed to operate the electric horn EXCEPT:

    A. Horn computer
    B. Horn relay
    C. Horn switch
    D. Clock spring

40. The vehicle climate control system creates passenger compartment heat by which method?

    A. An electric heater strip inside the dash
    B. A microwave heating grid inside the dash
    C. A heat exchanger, called a heater core, inside the dash
    D. A heat exchanger, called the evaporator, inside the dash.

41. Which of the following procedures would likely be performed during a typical 60,000-mile service?

    A. Replace the axle bearings.
    B. Drain and fill the transmission fluid.
    C. Replace the intake manifold gaskets.
    D. Drain and fill the windshield washer solvent.

42. Each of the following procedures would be performed during a typical oil change service EXCEPT:

    A. Drain and refill the engine oil.
    B. Replace the oil filter.
    C. Check the basic fluid levels and tire pressures.
    D. Replace the air filter.

43. All of the following are benefits of recommending additional services to a current customer who has left his/her car at your shop EXCEPT:

    A. The customer will begin to trust that the shop is looking out for his/her well-being and safety.
    B. The shop is more profitable from the increased sales of parts.
    C. The shop is more profitable from the increased sales of labor.
    D. The customer will compare the prices of the services with other shops.

44. Which items would be typically covered under a manufacturer's power train warranty?
    A. Belts and hoses
    B. Wheels and tires
    C. Front-end and suspension items
    D. Engine and transmission

45. Which of the following services would be included as part of a 30,000-mile service?
    A. Exhaust gasket replacement
    B. Intake manifold gasket replacement
    C. Fuel filter replacement
    D. Hub bearings repacking

46. Which of the following is the most likely location to find the vehicle production date?
    A. Stamped on the valve cover
    B. On the driver's side B-pillar
    C. The emission decal under the hood
    D. Inside the gas door

47. Service Consultant A says that when greeting a customer, you should offer your name and a handshake. Service Consultant B says that when greeting a customer, the service consultant should make eye contact and smile when welcoming them. Who is correct?
    A. A only
    B. B only
    C. Both A and B
    D. Neither A nor B

48. Service Consultant A says that some repair shops have to sublet body repairs to a body repair shop. Service Consultant B says that the charges for a sublet repair should be included on the repair order. Who is correct?
    A. A only
    B. B only
    C. Both A and B
    D. Neither A nor B

49. A customer arrives at the repair shop requesting an oil change and a tire rotation. Service Consultant A recommends that the shop perform a brake cleaning and inspection while the wheels are removed. Service Consultant B recommends that the fuel system be serviced while the vehicle is at the shop. Who is correct?
    A. A only
    B. B only
    C. Both A and B
    D. Neither A nor B

50. A vehicle that is in the shop for a cooling system repair needs to have a radiator tank replaced. Service Consultant A recommends that the radiator be replaced with a new unit since the shop is not equipped to make the radiator repair. Service Consultant B recommends sending the radiator to a radiator specialty shop to have the repair completed. Who is correct?
    A. A only
    B. B only
    C. Both A and B
    D. Neither A nor B

PREPARATION EXAM 4

1. Service Consultant A encourages his/her technicians to keep their work areas as clean and orderly as possible. Service Consultant B continually monitors the status of the repair work that is being performed and updates the customers if problems arise. Who is correct?
   A. A only
   B. B only
   C. Both A and B
   D. Neither A nor B

2. All of the following pieces of information are needed during the service write-up process EXCEPT:
   A. Driver's license number for the customer
   B. Vehicle year, make, and model
   C. Customer's name
   D. List of services to be addressed

3. Service Consultant A says a technical service bulletin (TSB) is issued when pattern failures happen to a certain type of vehicle. Service Consultant B says that vehicle manufacturers always cover the expenses associated with vehicles affected by TSBs. Who is correct?
   A. A only
   B. B only
   C. Both A and B
   D. Neither A nor B

4. Which of the following practices would be the LEAST LIKELY method for suitable alternative transportation?
   A. Explaining the location of the bus stop
   B. Taking the customer home
   C. Setting up a rental car
   D. Arranging a free loaner car

5. Service Consultant A says that the water pump is a component of the engine cooling system. Service Consultant B says the catalytic converter is a component of the emissions system. Who is correct?
   A. A only
   B. B only
   C. Both A and B
   D. Neither A nor B

6. Service Consultant A says that viewing the service history on a vehicle can show which technicians have performed repairs on a vehicle. Service Consultant B says that viewing the service history on a vehicle can show when the vehicle was delivered to the selling dealership. Who is correct?
   A. A only
   B. B only
   C. Both A and B
   D. Neither A nor B

7. A customer walks into the repair shop at 4:35 p.m. requesting an oil change and a cabin air filter replacement. The estimated time for this repair is 45 minutes and the shop closes at 5:00 p.m. Service Consultant A quickly completes the repair order and promises that the car will be completed that day before the close of business. Service Consultant B recommends that the customer reschedule this repair due to the limited time left in the day. Who is correct?
   A. A only
   B. B only
   C. Both A and B
   D. Neither A nor B

8. What is the most likely method to find an accurate repair procedure status?
   A. Ask the service manager to view the activity and progress.
   B. Walk through the shop to view the activity and progress.
   C. Ask the car porter to go inspect the technician's progress.
   D. Physically go and discuss the progress with the technician.

9. Which type of oil change reminder used on some late-model vehicles must be reset after each oil change?
   A. Oil pressure gauge
   B. Change oil indicator
   C. Oil pressure light
   D. Oil life chime

10. Service consultants need to present a professional image while at work. Service Consultant A wears flip-flop shoes with no socks to work. Service Consultant B wears a light blue uniform shirt with tan shorts to work. Who is correct?
   A. A only
   B. B only
   C. Both A and B
   D. Neither A nor B

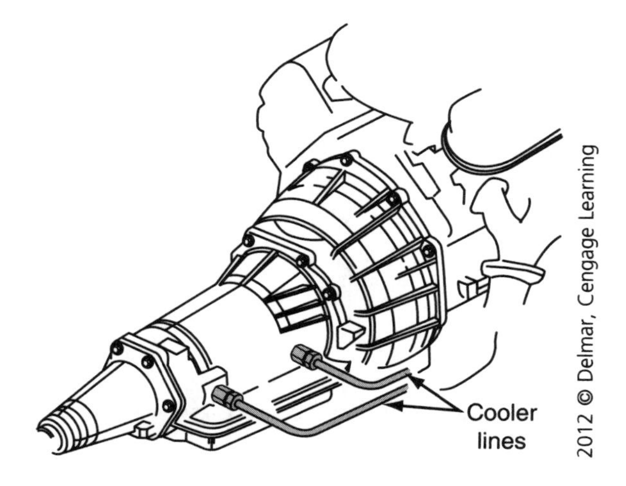

11. What purpose do the cooler lines in the figure above serve on an automatic transmission?
   A. To connect the transmission to the engine oil pan-mounted transmission cooler
   B. To connect the transmission to the radiator-mounted transmission cooler
   C. To connect the transmission coolant to the radiator
   D. To connect the transmission to the air-to-air transmission heater

12. A customer arrives to pick up his/her vehicle after the A/C compressor was replaced. After paying the bill, the customer comes back to the service desk to ask why the accumulator was replaced. Service Consultant A explains that all vehicles that have any type of A/C repairs need to have an accumulator replaced. Service Consultant B says that the accumulator should always be replaced whenever the compressor is replaced. Who is correct?

    A. A only
    B. B only
    C. Both A and B
    D. Neither A nor B

13. A vehicle is in the repair shop with a complaint of poor heater performance. Service Consultant A says that the engine cooling system may need to be diagnosed. Service Consultant B says that a stuck heater control valve could be the cause. Who is correct?

    A. A only
    B. B only
    C. Both A and B
    D. Neither A nor B

14. A vehicle with 115,250 miles on the odometer is in for service at a repair center for a problem of the heater not getting warm. The diagnosis is a restricted heater core. Service Consultant A prepares an estimate to replace the heater core. Service Consultant B prepares an estimate to replace the heater core, heater hoses, thermostat, and the radiator cap. Who is correct?

    A. A only
    B. B only
    C. Both A and B
    D. Neither A nor B

15. A vehicle is in the repair shop for a problem of the engine stalling when coming to a stop. The technician diagnoses the problem as a dirty throttle body and recommends a throttle-body cleaning service. Service Consultant A says that this repair will take several hours to complete. Service Consultant B says that this service operation involves removing the cylinder heads. Who is correct?

    A. A only
    B. B only
    C. Both A and B
    D. Neither A nor B

16. Which of the following car services is an operation that could be performed at the same time as an oil pan gasket replacement?

    A. Cabin air filter replacement
    B. Transmission fluid and filter service
    C. Intake manifold gasket
    D. Oil and filter change

17. Service Consultant A says that the airbag clock spring is located at the bottom of the steering column. Service Consultant B says that the airbag inflator module should be stored face down in a safe location while service is being performed. Who is correct?

    A. A only
    B. B only
    C. Both A and B
    D. Neither A nor B

18. Service Consultant A says that an accurate estimate is not important to have before calling the customer for approval. Service Consultant B says that the availability of the parts should be checked prior to calling the customer for approval. Who is correct?

    A. A only
    B. B only
    C. Both A and B
    D. Neither A nor B

19. Service Consultant A always provides clear estimates to her customers so they can make informed decisions about their vehicle. Service Consultant B always checks on the parts warranty in order to inform the customer about this detail. Who is correct?

    A. A only
    B. B only
    C. Both A and B
    D. Neither A nor B

20. Which of the following repair procedures would be the most important repair to be performed as related to vehicle safety?

    A. Wheel alignment
    B. Rear window defogger
    C. Brake line replacement
    D. Power window motor

21. Service Consultant A says that a three-year free replacement is a warranty feature on some premium batteries. Service Consultant B says that a lifetime guarantee is a warranty feature on some premium batteries. Who is correct?

    A. A only
    B. B only
    C. Both A and B
    D. Neither A nor B

22. Which of the following reasons could a service consultant use to finalize a service repair sale?

    A. The fuel economy will be reduced if the maintenance is not performed on time.
    B. The owner will be more likely to trade the vehicle because she is dissatisfied with it.
    C. The vehicle will be more reliable if the correct service and maintenance are performed on time.
    D. It is cheaper to drive a vehicle if maintenance is ignored.

23. All of the following methods of communicating a customer request to a technician are common EXCEPT:

    A. Paging the technician to the write-up desk to relay the message in person.

    B. Writing the message down and having the car porter give the message to the technician.

    C. Walking back to the technician and adding the customer request to the repair order.

    D. Having a car porter call the technician to relay the message.

24. A customer comments that his car pulls to the left while driving on a level road. Service Consultant A schedules the repair with the suspension specialist. Service Consultant B thinks that the problem could be unequal tire pressure and offers to check the tire pressure for free. Who is correct?

    A. A only

    B. B only

    C. Both A and B

    D. Neither A nor B

25. Which of the following activities would be the most likely method of promoting open communication among the repair shop employees?

    A. Weekly meetings that allow all employees to give input to the operation

    B. Putting a suggestion box in the employee break area

    C. Having a yearly Christmas party at a local restaurant

    D. Sending birthday cards to employees on their birthday

26. Service Consultant A always gives the highest priority to answering the phone rather than dealing with customers in the write-up area. Service Consultant B is able to carry on a phone conversation while dealing with a customer in person. Who is correct?

    A. A only

    B. B only

    C. Both A and B

    D. Neither A nor B

27. Service Consultant A says that the crankshaft sensor sends engine speed data to the engine computer. Service Consultant B says that the crankshaft sensor needs to be mounted near a reluctor ring. Who is correct?

    A. A only

    B. B only

    C. Both A and B

    D. Neither A nor B

28. Service Consultant A says that the alternator needs to be mounted near the flywheel. Service Consultant B says that the starter is driven by the accessory drive belt. Who is correct?

    A. A only

    B. B only

    C. Both A and B

    D. Neither A nor B

29. Which component is LEAST LIKELY to be needed in an overhead cam engine design?
    - A. Pushrod
    - B. Rocker arm
    - C. Intake valve
    - D. Exhaust valve

30. Service Consultant A promotes the repair shop by being knowledgeable about the certification status of the technicians who are employed there. Service Consultant B promotes the repair shop by being knowledgeable about the professionalism of the technicians who are employed there. Who is correct?
    - A. A only
    - B. B only
    - C. Both A and B
    - D. Neither A nor B

31. Which component would be most likely to be replaced during an intake manifold gasket replacement?
    - A. Rod bearing
    - B. Main bearing
    - C. Thermostat
    - D. Oil pump

32. Service Consultant A says that the clutch disc is located between the pressure plate and the flywheel. Service Consultant B says that the throw out bearing is located at the end of the crankshaft. Who is correct?
    - A. A only
    - B. B only
    - C. Both A and B
    - D. Neither A nor B

33. Which drive train component connects the transaxle to the drive hubs?
    - A. Hub bearing
    - B. Axle half shaft
    - C. Universal joint
    - D. Final drive assembly

34. A follow-up call is being completed. Service Consultant A asks the customer if he/she is pleased with the cleanliness of the repair shop. Service Consultant B asks the customer if he/she is pleased with the location of the shop. Who is correct?
    - A. A only
    - B. B only
    - C. Both A and B
    - D. Neither A nor B

35. All of the following components are parts of the steering system EXCEPT:

    A. Rack and pinion gear
    B. Strut
    C. Steering shaft
    D. Tie rod end

36. Which alignment angle is described as the inward- or outward-rolling direction of the tires?

    A. Toe
    B. Camber
    C. Caster
    D. Thrust angle

37. Which of the following anti-lock brake components is most likely to be located near each wheel?

    A. Modulator assembly
    B. Speed sensor
    C. Electronic control unit
    D. Deceleration sensor

38. Service Consultant A says that improved fuel economy would be a benefit of having well-maintained coolant. Service Consultant B says that improved vehicle handling would be a benefit of having a four-wheel alignment performed. Who is correct?

    A. A only
    B. B only
    C. Both A and B
    D. Neither A nor B

39. The HVAC system performs all of the following functions for a vehicle EXCEPT:

    A. Humidifies the air in dry conditions
    B. Heats the cabin when needed
    C. Cools the cabin when needed
    D. Dehumidifies the air when the A/C compressor is operated

40. A customer who has a failed engine that was obviously caused from the engine oil never being changed does not understand why his/her vehicle cannot to be repaired under warranty. Service Consultant A is very harsh in explaining that the customer should have maintained the vehicle by having the oil changed on a regular basis. Service Consultant B explains the situation to the customer by showing him/her how deteriorated the oil was. Who is correct?

    A. A only
    B. B only
    C. Both A and B
    D. Neither A nor B

41. Service Consultant A says that the throttle plates should be cleaned during a throttle body service. Service Consultant B says that some throttle bodies have to be removed from the engine to be cleaned. Who is correct?

A. A only
B. B only
C. Both A and B
D. Neither A nor B

42. A customer arrives at the repair shop to pick up a fleet vehicle after the shop has performed a warranty repair. Service Consultant A inquires about how many vehicles are in the fleet. Service Consultant B gives the customer a card and requests that the fleet manager call at the nearest convenient time to discuss future business. Who is correct?

A. A only
B. B only
C. Both A and B
D. Neither A nor B

43. Which is the LEAST LIKELY place to find a maintenance schedule for a vehicle?

A. Owner's manual
B. Warranty booklet
C. Electronic database
D. Shop manual

44. Which of the following examples describes how each customer should be greeted as he/she arrives at the repair facility?

A. Without emotion or any facial expression
B. With a slap on the back and an off-color joke
C. With a smile and a cordial greeting.
D. By sharing with them your daily problems in the service department

45. Service Consultant A says that items such as belts and hoses are considered wear items and are not typically covered by extended warranties. Service Consultant B says that major transmission components are considered power train items and would usually be covered by extended warranties. Who is correct?

A. A only
B. B only
C. Both A and B
D. Neither A nor B

46. Service Consultant A says that the tenth digit of the vehicle identification number (VIN) represents the engine. Service Consultant B says that the eighth digit of the VIN is the country of origin. Who is correct?

A. A only
B. B only
C. Both A and B
D. Neither A nor B

47. Service Consultant A says that a sedan is a car that has a retractable hardtop. Service Consultant B says that a coupe is a car that has four doors. Who is correct?

    A. A only
    B. B only
    C. Both A and B
    D. Neither A nor B

48. A customer is inquiring how often the tires should be rotated. Service Consultant A says the tires need to be rotated each time the oil is changed. Service Consultant B says the tires should be rotated at each 30,000-mile service. Who is correct?

    A. A only
    B. B only
    C. Both A and B
    D. Neither A nor B

49. Service Consultant A says that some repair shops choose to sublet radiator repairs to an outside facility. Service Consultant B says that some repair shops choose to sublet oil changes to an outside facility. Who is correct?

    A. A only
    B. B only
    C. Both A and B
    D. Neither A nor B

50. A customer calls and says that he nearly had an accident due to the engine stalling while crossing an intersection. The shop replaced a fuel pump three days ago. Service Consultant A recommends that the customer return to the service facility the next day to recheck the fuel system. Service Consultant B offers to send a wrecker to pick up the vehicle immediately due to the potentially unsafe conditions of the vehicle. Who is correct?

    A. A only
    B. B only
    C. Both A and B
    D. Neither A nor B

PREPARATION EXAM 5

1. A customer calls to speak with a service consultant who is already working with a customer. What should the service consultant taking the call do?

    A. Take the customer's name and number and assure them they will be called back.
    B. Transfer the customer to the parts department.
    C. Place the customer on hold until the consultant is available.
    D. Transfer the call to the service manager.

2. Technician efficiency needs to be monitored continuously. Service Consultant A monitors the weekly flat rate hours that each technician completes. Service Consultant B each technician on every job and compares this time with the flat rate time. Who is correct?

   A. A only

   B. B only

   C. Both A and B

   D. Neither A nor B

3. Which of the following terms is defined as a manufacturer's technical document created to assist technicians in repairing pattern failures more quickly?

   A. Service contract

   B. Campaign/recall

   C. Warranty

   D. Technical service bulletin (TSB)

4. All of the following are appropriate greeting approaches for a service consultant EXCEPT:

   A. Warm hug

   B. Friendly demeanor

   C. Handshake

   D. Smile

5. Which of the following components is integrated into the exhaust system?

   A. Knock sensor

   B. Camshaft sensor

   C. Crankshaft sensor

   D. Oxygen sensor

6. A customer's vehicle is being serviced for the first time by your shop. The customer states that she has had the same problem worked on at two other shops. Service Consultant A says that the technicians at this shop are much more competent than the other shops. Service Consultant B asks the customer to bring her repair records along to help the shop get an idea of the vehicle's history. Who is correct?

   A. A only

   B. B only

   C. Both A and B

   D. Neither A nor B

7. A customer comes in at 3 p.m. needing a complete brake job performed. The estimated time for this repair is four hours and the shop closes at 5 p.m. Service Consultant A recommends that the customer drop the car off and plan on picking the car up the next day at noon. Service Consultant B says that the shop can complete the job that day and encourages the technician to rush to complete the job on time. Who is correct?

   A. A only

   B. B only

   C. Both A and B

   D. Neither A nor B

8. A service consultant has prepared an estimate based upon a technician's diagnosis. Which of the following should the service consultant do first before providing the customer with the estimate?

   A. Check with the technician to plan a completion time.

   B. Check the availability of the repair parts.

   C. Test drive the vehicle.

   D. Identify additional needed services.

9. Service Consultant A says that the tire pressure warning light must be reset at each oil and filter change service. Service Consultant B says that the tire pressure warning light will sometimes illuminate as the weather gets colder. Who is correct?

   A. A only

   B. B only

   C. Both A and B

   D. Neither A nor B

10. A customer calls and states that his/her vehicle has an electrical problem that has been recurring after three attempts to repair it. Which of these should the service consultant do first?

    A. Check the repair history to see if this shop has worked on this vehicle's electrical problem before.

    B. Offer to diagnose the vehicle free of charge.

    C. Ask the customer to explain in detail all that has been done to the vehicle.

    D. Explain that some problems require several attempts to fix.

11. Service Consultant A says that the automatic transmission fluid should be changed more frequently if the vehicle is used to pull trailers with heavy loads. Service Consultant B says that all transmission fluid is the same as long as a name brand of fluid is used. Who is correct?

    A. A only

    B. B only

    C. Both A and B

    D. Neither A nor B

12. A customer is picking up his/her vehicle from a repair shop. Service Consultant A says that it is important to take the time to explain the work performed in as much detail as the customer requires. Service Consultant B says that if the customer asks questions it indicates he/she does not trust the shop. Who is correct?

    A. A only

    B. B only

    C. Both A and B

    D. Neither A nor B

13. Service Consultant A says that the starter relay is part of the starting system. Service Consultant B says that the voltage regulator is part of the starting system. Who is correct?

    A. A only

    B. B only

    C. Both A and B

    D. Neither A nor B

14. A vehicle is in the repair shop for an inspection. The technician recommends having the throttle body cleaned for a stalling problem and the valve cover gasket replaced for a small oil leak. The customer asks the service consultant to prioritize the two services in order of importance. Service Consultant A recommends the throttle body cleaning because the stalling problem is dangerous. Service Consultant B says that both items are of equal importance. Who is correct?
    A. A only
    B. B only
    C. Both A and B
    D. Neither A nor B

15. A technician turns in a repair order that recommends replacement of the constant velocity (CV) boot with no further description. Service Consultant A calls the customer to get approval to complete the repair. Service Consultant B consults with the technician to find out the reason for the replacement of the boot. Who is correct?
    A. A only
    B. B only
    C. Both A and B
    D. Neither A nor B

16. Service Consultant A says the benefit of a cabin air filter is that it reduces contaminants in the cabin area. Service Consultant B says that a benefit of a cabin air filter is cooler air from the vents. Who is correct?
    A. A only
    B. B only
    C. Both A and B
    D. Neither A nor B

17. Service Consultant A says that vehicles that do not have frames around the windows are known as hard tops. Service Consultant B says that a hatchback is a vehicle with a rear trunk/window combination that lifts up. Who is correct?
    A. A only
    B. B only
    C. Both A and B
    D. Neither A nor B

18. Service Consultant A says that providing an estimate is required by law in all states. Service Consultant B says that explaining the details of the estimate will give customers a better understanding of what is going to be done to their vehicle. Who is correct?
    A. A only
    B. B only
    C. Both A and B
    D. Neither A nor B

19. Service Consultant A provides clear and understandable answers to detailed questions that the customers have. Service Consultant B shows customers the details of the repair order before completing the repair order. Who is correct?
    A. A only
    B. B only
    C. Both A and B
    D. Neither A nor B

20. A vehicle is in the repair shop for a water pump replacement. Service Consultant A recommends that the cooling system be totally flushed and refilled with new coolant while the water pump is being replaced. Service Consultant B recommends that the A/C system should be recovered and recharged while the water pump is being replaced. Who is correct?

    A. A only
    B. B only
    C. Both A and B
    D. Neither A nor B

21. Service Consultant A says that a benefit to having regular brake inspections is reliable stopping operation of the vehicle. Service Consultant B says that a benefit to regular brake inspections is improved fuel economy. Who is correct?

    A. A only
    B. B only
    C. Both A and B
    D. Neither A nor B

22. Service Consultant A says that an accurate estimate is NOT critical to have before calling the customer for approval. Service Consultant B says that an exact completion time is important to have before calling for customer approval. Who is correct?

    A. A only
    B. B only
    C. Both A and B
    D. Neither A nor B

23. A customer has just given approval for repair of his/her vehicle. Service Consultant A says the technician should be provided with the approved work order. Service Consultant B says documentation of the customer's approval should be on the work order. Who is correct?

    A. A only
    B. B only
    C. Both A and B
    D. Neither A nor B

24. Recommending additional services is a common practice for service consultants. Service Consultant A says that the shop will be more profitable due to increased labor charges if additional services are sold. Service Consultant B says that customers will appreciate the fact that the technicians are inspecting their vehicles for safety and maintenance items. Who is correct?

    A. A only
    B. B only
    C. Both A and B
    D. Neither A nor B

25. Which is the LEAST LIKELY method to be used when checking for quality control in a repair shop?

    A. Keep a log of repeat repairs for each technician.
    B. Keep a total of labor "write-off" amounts for each technician.
    C. Keep a log of redeemed coupons for each month.
    D. Keep a log of repeat repairs for the whole shop.

26. A service consultant has just completed compiling and writing up a customer's concerns. Which of the following should she do next?

    A. Confirm the accuracy of the information on the repair order.
    B. Arrange a ride home for the customer.
    C. Offer an estimate for the repairs.
    D. Have the porter wash the car.

27. Which engine component regulates engine temperature?

    A. Heater core
    B. Radiator
    C. Water pump
    D. Thermostat

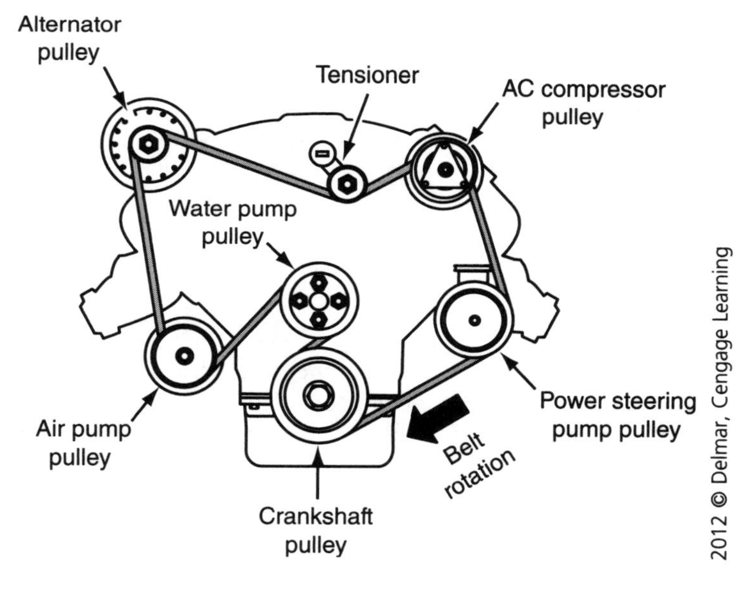

28. Referring to the figure above, Service Consultant A says that the alternator provides the belt tension on the drive belt. Service Consultant B says that the belt used in the figure is a V-belt. Who is correct?

    A. A only
    B. B only
    C. Both A and B
    D. Neither A nor B

29. Which of the following components is part of a typical charging system?

    A. Starter solenoid
    B. Park/neutral switch
    C. Ignition switch
    D. Voltage regulator

30. Service Consultant A suggests that offering a customer a ride home or to work represents alternative transportation. Service Consultant B suggests that driving the customer to the bus stop is providing alternative transportation. Who is correct?

    A. A only
    B. B only
    C. Both A and B
    D. Neither A nor B

31. Service Consultant A says that the radiator is a component of the engine cooling system. Service Consultant B says the catalytic converter is a component of the ignition system. Who is correct?

    A. A only
    B. B only
    C. Both A and B
    D. Neither A nor B

32. Service Consultant A says that rear-wheel drive vehicles use a half shaft to connect the transmission to the rear axle. Service Consultant B says that front-wheel drive vehicles use a driveshaft with universal joints to connect the transaxle to the drive wheels. Who is correct?

    A. A only
    B. B only
    C. Both A and B
    D. Neither A nor B

33. Service Consultant A says that the engine flywheel drives the clutch disc when the clutch pedal is released. Service Consultant B says that the pressure plate is the part of the system that compresses the clutch disc into the flywheel when the clutch pedal is depressed. Who is correct?

    A. A only
    B. B only
    C. Both A and B
    D. Neither A nor B

34. Which of the following service consultant choices for appearance would be LEAST LIKELY to present a professional image?

    A. Green shirt and blue pants
    B. Blue jeans and tennis shoes
    C. White shirt and tan pants
    D. Clean and neatly groomed hair

35. Which of the following components is part of the antilock brake system (ABS)?
    A. Master cylinder
    B. Wheel cylinder
    C. Caliper
    D. Hydraulic control unit

36. Which component of the steering system connects the steering gear linkage to the steering knuckle?
    A. Spring
    B. Tie rod end
    C. Strut
    D. Control arm

37. Service Consultant A does not recommend that the alignment be checked prior to having new tires installed. Service Consultant B quotes two levels of tires in order to let the customer decide the expense. Who is correct?
    A. A only
    B. B only
    C. Both A and B
    D. Neither A nor B

38. When a customer objects to the cost of a given repair, the best response by the service consultant would be to:
    A. Refer the customer to the service manager.
    B. Offer the customer a discount to encourage his/her approval.
    C. Explain the benefits of having the repair performed.
    D. Reschedule the repair for a different time.

39. Service Consultant A says that the evaporator core is the source of heat in the HVAC system. Service Consultant B says that the blend door is a device that controls the temperature of the air that is discharged from the HVAC system? Who is correct?
    A. A only
    B. B only
    C. Both A and B
    D. Neither A nor B

40. Service Consultant A adds the labor total when calculating an estimate for a repair. Service Consultant B adds the parts total when calculating an estimate for a repair. Who is correct?
    A. A only
    B. B only
    C. Both A and B
    D. Neither A nor B

41. All of the following procedures are performed during a typical 90,000-mile service EXCEPT:
    A. Replace the spark plugs.
    B. Replace the fuel filter.
    C. Inspect and adjust the brake system.
    D. Replace the A/C compressor.

42. What is the LEAST LIKELY step that would be followed by a service consultant to verify the accuracy of a repair order prior to calling the customer for repair authorization?

    A. Add the labor and the parts up in your mind.
    B. Input all the service repair-related data into the computer and verify the availability of the parts.
    C. Add the cost of the parts, the cost of the labor, the tax, and the cost of sublet repairs.
    D. Input the parts cost, labor cost, tax cost, and the sublet repair cost into a calculator.

43. Service Consultant A says that maintenance schedules are printed in the vehicle owner's manual. Service Consultant B says that maintenance schedules are selected based on the customer's use of the vehicle. Who is correct?

    A. A only
    B. B only
    C. Both A and B
    D. Neither A nor B

44. All of the following items are needed when collecting vehicle information for the vehicle repair order EXCEPT:

    A. Mileage
    B. Interior color
    C. Vehicle model
    D. Vehicle make

45. What is the most likely repair that would be covered by a service contract?

    A. Brake pads
    B. Wiper blade
    C. Water pump
    D. Air filter

46. Which VIN digit represents the vehicle year?

    A. First
    B. Sixth
    C. Eighth
    D. Tenth

47. Service Consultant A uses the internet to locate pictures of the major vehicle systems in order to show the customer which items need service. Service Consultant B keeps a notebook with various pictures of vehicle systems at the service desk to show the customer which items need service. Who is correct?

    A. A only
    B. B only
    C. Both A and B
    D. Neither A nor B

48. Service Consultant A says that it is very helpful to know what has been done to the vehicle in the past. Service Consultant B says that some shops have computerized software that stores the service data on each vehicle that has been in the repair facility. Who is correct?
   A. A only
   B. B only
   C. Both A and B
   D. Neither A nor B

49. Service Consultant A says that some shops sublet body repairs to a body shop. Service Consultant B says that sublet repairs should be added to the repair order when adding up the total bill for the customer. Who is correct?
   A. A only
   B. B only
   C. Both A and B
   D. Neither A nor B

50. An upset customer comes in when the service department is very busy with a complaint about past repair service. Service Consultant A listens to the customer until he has the opportunity to show that he/she is wrong. Service Consultant B lets the customer tell the whole story. Who is correct?
   A. A only
   B. B only
   C. Both A and B
   D. Neither A nor B

PREPARATION EXAM 6

1. A potential customer calls very concerned about an estimate received from another shop. Which of the following should the service consultant do?
   A. Show concern for the potential customer and offer an appointment for a second opinion.
   B. Research the amount that service consultant's shop would charge for that service.
   C. Offer a discount if the potential customer brings the vehicle in.
   D. Tell the potential customer that the other shop is dishonest.

2. Which fuel system component is mounted in the fuel tank?
   A. Fuel pump
   B. Fuel supply line
   C. Fuel injector
   D. Fuel level gauge

3. Which of the following items would be typically covered under a manufacturer's power train warranty?
   A. Tires
   B. Transmission
   C. Brake pads
   D. Fuel filter

4. Service Consultant A says that customers expect to be treated with respect when they arrive at the repair shop. Service Consultant B says that customers expect a sincere service-oriented attitude when they are doing business with a repair shop. Who is correct?

   A. A only
   B. B only
   C. Both A and B
   D. Neither A nor B

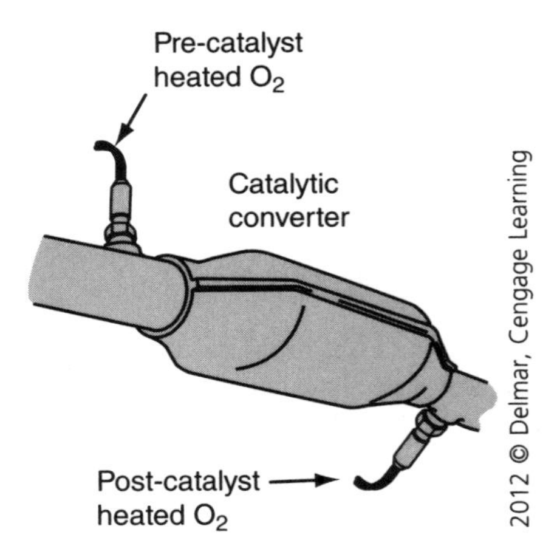

5. Referring to the figure above, Service Consultant A says that this vehicle uses an OBD-II emission system. Service Consultant B says that the post-catalyst O_2 sensor measures the efficiency of the catalytic converter. Who is correct?

   A. A only
   B. B only
   C. Both A and B
   D. Neither A nor B

6. Which of the following is most likely to have the greatest positive impact on gaining repeat customers for a repair shop?

   A. Receiving an honest and fair service at a reasonable price
   B. The service consultant's appearance
   C. Printed advertisement in a local newspaper
   D. Discount pricing

7. A customer walks into the repair shop at 4:30 p.m. requesting an oil change and a tire rotation. The estimated time for this repair is 45 minutes and the shop closes at 5:30 p.m. A service technician is available to perform this service. Service Consultant A quickly completes the repair order and promises that the car will be completed that same day before the close of business. Service Consultant B recommends that the customer reschedule this repair due to the limited time left in the day. Who is correct?

   A. A only
   B. B only
   C. Both A and B
   D. Neither A nor B

8. A vehicle in the shop for an oil change shows approximately 59,000 miles on the odometer. What should the service consultant do?

   A. Advise the customer that the 60,000-mile service is covered under the power train warranty.
   B. Quote a price and offer to schedule an appointment for a 60,000-mile service.
   C. Offer a discount if the customer schedules the appointment within the next week.
   D. Advise the customer that the 60,000-mile service is critical and the vehicle is not safe to drive if the service is not performed.

9. Which of the following conditions will cause the tire pressure warning light to illuminate?

   A. Tire pressure that is three to six pounds above the specification
   B. Tires that are worn below 4/32-inch tread depth
   C. Tire pressure that drops four to six pounds below the specification
   D. Tires that are out of balance

10. Service Consultant A wears his/her uniform without the shirt tucked in. Service Consultant B wears a casual uniform to work. Who is correct?

   A. A only
   B. B only
   C. Both A and B
   D. Neither A nor B

11. Which engine component rotates and causes the valve to open at the correct time?

   A. Crankshaft
   B. Timing gear
   C. Camshaft
   D. Flywheel

12. Service Consultant A adds the miscellaneous expense total when calculating an estimate for a repair. Service Consultant B adds the sublet repair totals when calculating an estimate for a repair. Who is correct?

   A. A only
   B. B only
   C. Both A and B
   D. Neither A nor B

13. Service Consultant A says that the starter relay is part of the charging system. Service Consultant B says that the voltage regulator is part of the charging system. Who is correct?

   A. A only
   B. B only
   C. Both A and B
   D. Neither A nor B

14. Service Consultant A refers the customer to the service manager if the customer asks detailed questions about why a particular service is being recommended. Service Consultant B provides clear and understandable answers to detailed questions that his customers have. Who is correct?

    A. A only
    B. B only
    C. Both A and B
    D. Neither A nor B

15. Service Consultant A monitors the weekly flat-rate hours that each technician flags to determine the efficiency of the technicians. Service Consultant B monitors the total shop hours for each month to determine shop productivity. Who is correct?

    A. A only
    B. B only
    C. Both A and B
    D. Neither A nor B

16. Service Consultant A says that the benefit of having regular oil changes is reduced engine wear. Service Consultant B says the benefit of replacing the air filter at regular intervals is reduced engine wear. Who is correct?

    A. A only
    B. B only
    C. Both A and B
    D. Neither A nor B

17. Service Consultant A says that the negative battery cable connects to the engine block. Service Consultant B says the positive battery cable connects to the starter solenoid. Who is correct?

    A. A only
    B. B only
    C. Both A and B
    D. Neither A nor B

18. Which of the following repairs would be the highest priority for a customer to repair in relation to the safety of the vehicle?

    A. Replacing the power window motor
    B. Replacing the blower motor
    C. Replacing the radio
    D. Replacing the clock spring for the airbag inflator

19. A customer brings a car to the repair shop with a complaint that the steering wheel vibrates while braking. Service Consultant A recommends having the brake technician diagnose the problem. Service Consultant B recommends that the rotors will need to be turned or replaced. Who is correct?

    A. A only
    B. B only
    C. Both A and B
    D. Neither A nor B

20. A customer arrives at the repair shop requesting an oil change and a tire rotation. Service Consultant A recommends that the shop perform a brake cleaning and inspection while the wheels are removed. Service Consultant B recommends that the air filter be inspected while the vehicle is at the shop. Who is correct?

    A. A only
    B. B only
    C. Both A and B
    D. Neither A nor B

21. Service Consultant A says that providing an accurate estimate is a useful tool for closing a sale. Service Consultant B says that explaining the warranty on the repair parts is a good way to close a sale. Who is correct?

    A. A only
    B. B only
    C. Both A and B
    D. Neither A nor B

22. A technician turns in a repair order that recommends replacement of the fuel filter. Service Consultant A calls the customer to get approval to complete the repair. Service Consultant B consults with the technician to find out the reason for filter replacement. Who is correct?

    A. A only
    B. B only
    C. Both A and B
    D. Neither A nor B

23. Which of the following is the most likely method to be used when checking for quality control in a repair shop?

    A. Keep a record of repeat repairs for each technician.
    B. Monitor the total shop revenue each month.
    C. Keep a record of redeemed coupons for each month.
    D. Keep a record of random maintenance customers for each month.

24. All of the following are benefits of recommending overdue services to a current customer who has left his/her car at your shop EXCEPT:

    A. The customer will begin to trust that the shop is looking out for his/her well-being and safety.
    B. The employees will have to work late to complete the repair.
    C. The shop is more profitable from the increased sale of labor.
    D. The shop is more profitable from the increased sale of parts.

25. Service Consultant A meets with the service manager each week to discuss the stresses involved with the job. Service Consultant B continuously monitors the shop environment by discussing the daily activities with the shop technicians. Who is correct?

    A. A only
    B. B only
    C. Both A and B
    D. Neither A nor B

26. When writing up a customer's work order, which of the following is the first thing to ask for?

    A. The vehicle identification number (VIN)
    B. The customer's name
    C. The main customer concern
    D. The license plate number

27. Service Consultant A says that the crankshaft sensor sends engine speed data to the electronic control module (ECM). Service Consultant B says that the crankshaft sensor needs to be mounted near the drive belt. Who is correct?

    A. A only
    B. B only
    C. Both A and B
    D. Neither A nor B

28. Service Consultant A says that the alternator maintains belt tension on a serpentine belt system. Service Consultant B says that a serpentine belt is wider than a V-belt. Who is correct?

    A. A only
    B. B only
    C. Both A and B
    D. Neither A nor B

29. All of the following components are emissions system-related devices EXCEPT:

    A. Knock sensor
    B. Cabin temperature sensor
    C. Crankshaft sensor
    D. Oxygen sensor

30. Service Consultant A says that the some customers will ask for a ride back to their house when they make their appointment. Service Consultant B says that some customers will ask for a ride back to their work location when they make their appointment. Who is correct?

    A. A only
    B. B only
    C. Both A and B
    D. Neither A nor B

31. Service Consultant A says that rear-wheel drive vehicles use a driveshaft with universal joints on each end to connect the transmission to the rear axle. Service Consultant B says that front-wheel drive vehicles use a half shaft with constant velocity (CV) joints to connect the transaxle to the drive wheels. Who is correct?

    A. A only
    B. B only
    C. Both A and B
    D. Neither A nor B

32. Service Consultant A says that the clutch disc is splined onto the transmission input shaft. Service Consultant B says that the pressure plate is the part of the system that compresses the clutch disc into the flywheel when the clutch pedal is released. Who is correct?

 A. A only
 B. B only
 C. Both A and B
 D. Neither A nor B

33. Service Consultant A says that the automatic transmission fluid should be changed more frequently if the vehicle is driven at highway speeds most of the time. Service Consultant B says that the transmission fluid is cooled by routing it into the transmission oil cooler, which is mounted inside the radiator. Who is correct?

 A. A only
 B. B only
 C. Both A and B
 D. Neither A nor B

34. A customer arrives to pick up his vehicle after the transmission has been replaced. Service Consultant A carefully explains all of the repairs and charges and asks the customer if there are any questions. Service Consultant B calls the service porter to pull the vehicle around to the pickup area after answering all of the customer's questions. Who is correct?

 A. A only
 B. B only
 C. Both A and B
 D. Neither A nor B

35. What is the most likely location for the antilock brake system (ABS) hydraulic modulator?

 A. Near the wheel speed sensor
 B. Inside the engine compartment
 C. Inside the passenger compartment
 D. Next to the fuel tank

36. Which component of the suspension system connects the steering knuckle to the vehicle body?

 A. Spring
 B. Tie rod end
 C. Strut
 D. Control arm

37. Service Consultant A recommends running the tire pressure 15 psi higher than the specification in order to increase tire life. Service Consultant B recommends rotating the tires every 7,500 miles in order to increase tire life. Who is correct?

 A. A only
 B. B only
 C. Both A and B
 D. Neither A nor B

38. A customer is objecting to the cost of the repair bill when picking up her vehicle. The repair had been authorized by her husband prior to completing the repair. Service Consultant A asks the customer to call her husband to verify his authorization. Service Consultant B refers the customer to the service manager to find a solution. Who is correct?

    A. A only
    B. B only
    C. Both A and B
    D. Neither A nor B

39. Service Consultant A says that the heater core is the source of heat in the HVAC system. Service Consultant B says that the blend door is the device that controls the location to which the air is discharged in the HVAC system? Who is correct?

    A. A only
    B. B only
    C. Both A and B
    D. Neither A nor B

40. Which of the following repair procedures would be the LEAST LIKELY to be considered a high priority repair?

    A. Replacement of a worn tie rod end
    B. Brake fluid flush and fill
    C. Replacement of a tire with the steel showing
    D. Replacement of worn front brake pads

41. All of the following procedures would be completed during a 90,000-mile service EXCEPT:

    A. Drain and fill the automatic transmission fluid.
    B. Replace the spark plugs.
    C. Replace the alternator.
    D. Replace the fuel filter.

42. Service Consultant A shares the expected completion time with the technician so that completion expectations are clear. Service Consultant B asks the technician to alert him if the completion time changes due to a problem in the repair. Who is correct?

    A. A only
    B. B only
    C. Both A and B
    D. Neither A nor B

43. Service Consultant A says that maintenance schedules are printed in the vehicle owner's manual. Service Consultant B says that maintenance schedules are the same for all types of driving styles. Who is correct?

    A. A only
    B. B only
    C. Both A and B
    D. Neither A nor B

44. All of the following pieces of customer information might be included on a repair order EXCEPT:
    A. Customer's email address
    B. Vehicle make and model
    C. Cell phone number
    D. Service consultant's cell phone number

45. Service Consultant A says that a manufacturer technical service bulletin (TSB) is a technical document created to help technicians repair pattern failures more quickly. Service Consultant B says that a power train warranty is a program that manufacturers create to invite customers to bring their vehicles back in for a free repair in order to correct a safety fault in the vehicle. Who is correct?
    A. A only
    B. B only
    C. Both A and B
    D. Neither A nor B

46. Which VIN digit represents the vehicle engine?
    A. First
    B. Sixth
    C. Eighth
    D. Tenth

47. Service Consultant A says that the vehicle identification number has a digit that reveals the model of the vehicle. Service Consultant B says that a sedan is a car that has only two doors. Who is correct?
    A. A only
    B. B only
    C. Both A and B
    D. Neither A nor B

48. What is the most likely reason to suggest additional repair work on a customer's vehicle?
    A. The responsibility of the repair shop to advise the customer of needed service
    B. To increase the amount of labor charges for the shop
    C. To increase the amount of parts sold by the shop
    D. To help customers maintain vehicle value by having needed repair work done

49. Service Consultant A schedules appointments over the phone and sometimes forgets to add them to the appointment log. Service Consultant B invites customers that he meets after work to come to the repair shop without adding them to the appointment log. Who is correct?
    A. A only
    B. B only
    C. Both A and B
    D. Neither A nor B

50. An upset customer calls and expresses frustration because he nearly had an accident due to the brake pedal going all the way to the floor. The shop had performed a complete brake job the day before. Service Consultant A recommends that the customer return to the service facility in two days for a recheck of the brake system. Service Consultant B offers to send a wrecker to pick up the vehicle immediately due to the potentially unsafe condition of the vehicle. Who is correct?
    A. A only
    B. B only
    C. Both A and B
    D. Neither A nor B

SECTION

6 Answer Keys and Explanations

INTRODUCTION

Included in this section are the answer keys for each preparation exam, followed by individual, detailed answer explanations and a reference identifying the designated task area being assessed by each specific question. This additional reference information may prove useful if you need to refer back to the task list located in Section 4 of this book for additional support.

PREPARATION EXAM 1—ANSWER KEY

1. C
2. B
3. C
4. B
5. B
6. C
7. C
8. C
9. A
10. C
11. A
12. B
13. C
14. A
15. B
16. B
17. C
18. B
19. B
20. B
21. C
22. C
23. A
24. D
25. B
26. A
27. C
28. D
29. B
30. A
31. D
32. C
33. A
34. B
35. B
36. C
37. C
38. C
39. A
40. D
41. C
42. C
43. B
44. A
45. C
46. A
47. B
48. C
49. C
50. A

PREPARATION EXAM 1—EXPLANATIONS

1. Service Consultant A speaks clearly when having a conversation with a customer on the phone. Service Consultant B says that treating customers with dignity and respect on the phone is a positive business trait. Who is correct?

TASK A.1.1

A. A only

B. B only

C. Both A and B

D. Neither A nor B

Answer A is incorrect. Service Consultant B is also correct.

Answer B is incorrect. Service Consultant A is also correct.

Answer C is correct. Both Service Consultants are correct. Speaking clearly and treating customers with dignity and respect are both very positive methods to use when communicating with customers. All customers should be treated like valuable assets to a service repair business.

Answer D is incorrect. Both Service Consultants are correct.

2. Service Consultant A asks the customer to speak very slowly to allow every comment to be written on the repair order. Service Consultant B asks open-ended questions when attempting to identify the customer concern to be written on the repair order. Who is correct?

TASK A.1.3

A. A only

B. B only

C. Both A and B

D. Neither A nor B

Answer A is incorrect. The service consultant does not need to write every small detail that a customer says on the repair order. It is wise to listen to the whole conversation and then try to summarize what the customer is saying. After putting these details down, it should be read back to the customer to verify the accuracy.

Answer B is correct. Only Service Consultant B is correct. Open-ended questions work well when trying to get the big picture of what the customer concern really is. After listening to the customer answer these questions, the service consultant should summarize the details on the repair order.

Answer C is incorrect. Only Service Consultant B is correct.

Answer D is incorrect. Service Consultant B is correct.

3. Service consultants should use all of the following when greeting a new customer EXCEPT:

TASK A.1.4

A. Good eye contact

B. Cordial handshake

C. Customer's first name

D. Genuine smile

Answer A is incorrect. Using good eye contact is always recommended when dealing with service repair customers.

Answer B is incorrect. A cordial handshake is a recommended greeting for a new customer.

Answer C is correct. It is recommended to use the customer's last name when dealing with them for the first time. For example, Mike Smith should be addressed as "Mr. Smith" to show due respect to the clients of the repair shop.

Answer D is incorrect. A genuine smile should be a part of greeting all customers in a repair shop.

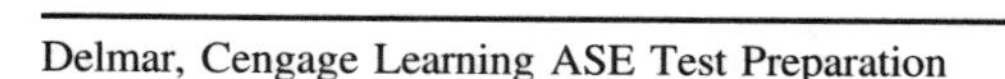

TASK A.1.7

4. Which of the following pieces of data can be determined from viewing the service history of a vehicle?

   A. Vehicle production date
   B. Service repair procedures performed at this location
   C. Location of the selling dealer
   D. Open recalls for the vehicle

Answer A is incorrect. The production date can be found on the B-pillar on the driver's side or retrieved by a dealer from the dealer service network.

Answer B is correct. The service history shows a list of the repair procedures performed at that service location. This information is helpful when planning current repair and maintenance for each vehicle.

Answer C is incorrect. The selling location for a vehicle would likely be available through a dealer service network inquiry.

Answer D is incorrect. A list of open recalls for a vehicle would likely be available through a dealer service network inquiry.

TASK A.1.9

5. A customer has an electrical problem with a vehicle. Service Consultant A promises a completion time to the customer during the write-up process. Service Consultant B gives the customer regular updates on the status of his/her vehicle throughout the repair visit. Who is correct?

   A. A only
   B. B only
   C. Both A and B
   D. Neither A nor B

Answer A is incorrect. It is not a good idea to give completion times to the customer during the write-up process on a problem that involves diagnosis. It is very challenging to predict the completion time before knowing what is causing the problem.

Answer B is correct. Only Service Consultant B is correct. Keeping the customer informed throughout the repair process is highly recommended.

Answer C is incorrect. Only Service Consultant B is correct.

Answer D is incorrect. Service Consultant B is correct.

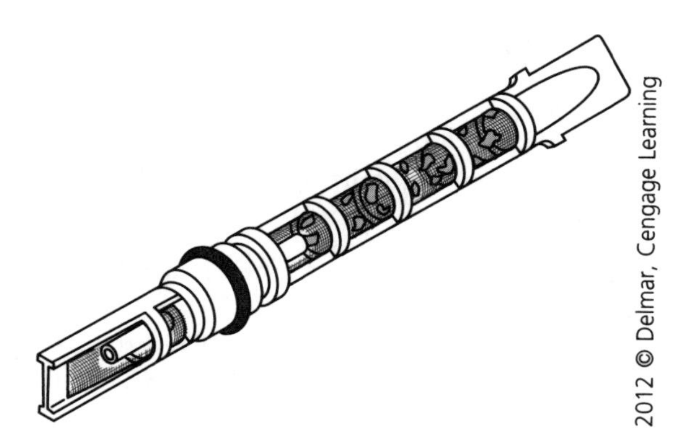

6. Service Consultant A says that the component shown in the figure above needs to be serviced and that the A/C system from which it came needs to be flushed. Service Consultant B says that a failed A/C compressor could have caused the debris on the screen. Who is correct?

TASK B.4.1

A. A only
B. B only
C. Both A and B
D. Neither A nor B

Answer A is incorrect. Service Consultant B is also correct.

Answer B is incorrect. Service Consultant A is also correct.

Answer C is correct. Both Service Consultants are correct. The orifice tube in the picture is very contaminated and would need to be replaced. In addition, the A/C system would need to be flushed to remove the contaminants from the rest of the system. A failed A/C compressor is one of the potential causes for the contaminated orifice tube.

Answer D is incorrect. Both Service Consultants are correct.

7. Service Consultant A says that when writing up a comeback/warranty ticket, it is necessary to review previous repair orders with the customer. Service Consultant B says that when writing up a comeback/warranty ticket, it is necessary to ask the customer to restate the symptoms he is experiencing. Who is right?

TASK A.1.2, A.1.7

A. A only
B. B only
C. Both A and B
D. Neither A nor B

Answer A is incorrect. Service Consultant B is also correct.

Answer B is incorrect. Service Consultant A is also correct.

Answer C is correct. Both Service Consultants are correct. Some customers might give the impression that the facility has serviced the vehicle before, but upon service history review, it may be discovered rather that another facility has looked at the vehicle before for the problem. Often, the customer may provide new information that will help solve the problem.

Answer D is incorrect. Both Service Consultants are correct. A check of history combined with new information may result in a fix.

TASK A.1.13

8. All of the following are positive results from a customer follow-up call EXCEPT:

   A. Constructive criticism is gained from customer comments.
   B. Shop income can be increased due to new appointments.
   C. The service consultant can confront the customer who has negative feedback.
   D. The customer will appreciate the follow-up call.

Answer A is incorrect. Obtaining constructive feedback from customers is an effective way to make improvements to a service facility.

Answer B is incorrect. Performing customer follow-up calls can potentially increase shop income. One way this can happen is that the customer may recognize it is time for more service and schedule that during the call. Shop income can also potentially rise due to customers sharing the details with their family and friends about the shop's conscientious calls to check on their vehicle.

Answer C is correct. A service consultant should not be confrontational when performing a follow-up call. If a customer is not happy with a service that they received, the service consultant should invite the customer back to try to remedy the grievance.

Answer D is incorrect. Customers will appreciate that the service facility is requesting feedback from their experience at the service facility.

TASK B.6.1, C.4

9. A vehicle had a new water pump installed just 30 days ago and is back in the repair shop with the water pump bearing making a growling noise. Service Consultant A recognizes that this condition should be covered under the parts warranty of the previously installed water pump. Service Consultant B says that the customer should have to pay for this repair again because engine parts do not carry any warranty coverage. Who is correct?

   A. A only
   B. B only
   C. Both A and B
   D. Neither A nor B

Answer A is correct. Only Service Consultant A is correct. This is an obvious case of a parts warranty. The Service Consultant should not charge the customer for this repair. The manufacturer of the water pump should pay this claim.

Answer B is incorrect. Nearly all vehicle parts have some warranty for some length of time.

Answer C is incorrect. Only Service Consultant A is correct.

Answer D is incorrect. Service Consultant A is correct.

TASK B.7.4

10. Service Consultant A says that a hard-top convertible will have limited trunk space when the top is down due to the vehicle design. Service Consultant B says that a soft-top convertible will typically have a more rigid body and frame due to the absence of the structural top. Who is correct?

   A. A only
   B. B only
   C. Both A and B
   D. Neither A nor B

Answer A is incorrect. Service Consultant B is also correct.

Answer B is incorrect. Service Consultant A is also correct.

Answer C is correct. Both Service Consultants are correct. A hard-top convertible will have limited trunk space when the top is down due to the vehicle design. The folding hard top mechanism collapses into the trunk area when the top is down. A soft-top convertible will be built with more rigid body and frame components to make up for the loss of a structural top.

Answer D is incorrect. Both Service Consultants are correct.

11. Service Consultant A says that major engine components are typically covered by extended warranties. Service Consultant B says that brake pads are typically covered by extended warranties. Who is correct?

TASK B.6.1, B.6.2

A. A only

B. B only

C. Both A and B

D. Neither A nor B

Answer A is correct. Only Service Consultant A is correct. Major power train components are usually covered by extended warranties. A service consultant needs to read the fine print carefully when dealing with these types of policies. It is typically necessary to call the company to get approval for the repair before completing the repair.

Answer B is incorrect. Brake pads are typically considered maintenance items and are not covered by most extended warranties.

Answer C is incorrect. Only Service Consultant A is correct.

Answer D is incorrect. Service Consultant A is correct.

12. Service Consultant A says that giving a ballpark estimate during the write-up process is a good idea. Service Consultant B says that repair estimates should only be given after the technician has diagnosed the vehicle and checked the cost of the parts needed. Who is correct?

TASK A.2.1, A.3.3

A. A only

B. B only

C. Both A and B

D. Neither A nor B

Answer A is incorrect. Giving any type of total estimate during the write-up process is not advisable. However, it is sometimes necessary to give an estimate of diagnostic time when the customer requests it. For example, the customer may only be willing to invest an hour or two of labor into diagnosing their problem. In these cases, the technician should be advised to not go over the approved diagnostic time.

Answer B is correct. Only Service Consultant B is correct. Total cost repair estimates should only be given after the vehicle has been diagnosed. Only then can the total parts and labor be added to accurately predict what the cost of the repair will be.

Answer C is incorrect. Only Service Consultant B is correct.

Answer D is incorrect. Service Consultant B is correct.

13. Service Consultant A recommends that the vehicle tires be rotated at every other oil change. Service Consultant B recommends that the vehicle tire pressure be checked at every oil change. Who is correct?

TASK B.3.3

A. A only

B. B only

C. Both A and B

D. Neither A nor B

Answer A is incorrect. Service Consultant B is also correct.

Answer B is incorrect. Service Consultant A is also correct.

Answer C is correct. Both Service Consultants are correct. A typical tire rotation schedule would be about every 7,500 miles, which is approximately every other oil change. The tire pressure should be checked at every oil change.

Answer D is incorrect. Both Service Consultants are correct.

TASK B.1.2

14. Service Consultant A says that some engines use a timing chain to connect the crankshaft to the camshaft. Service Consultant B says that the crankshaft and camshaft turn at the same speed when the engine is running. Who is correct?

A. A only
B. B only
C. Both A and B
D. Neither A nor B

Answer A is correct. Only Service Consultant A is correct. Some engines use a timing chain to connect the camshaft and crankshaft. Other types of timing mechanisms include timing belts and timing gears.

Answer B is incorrect. The crankshaft turns at twice the speed of the camshaft.

Answer C is incorrect. Only Service Consultant A is correct.

Answer D is incorrect. Service Consultant A is correct.

TASK A.1.12

15. Service Consultant A wears jeans and a worn shirt to work. Service Consultant B wears clean "business casual" clothes and always tucks in his/her shirt while at work. Who is correct?

A. A only
B. B only
C. Both A and B
D. Neither A nor B

Answer A is incorrect. Wearing blue jeans and a worn shirt does not present a professional image to a customer.

Answer B is correct. Only Service Consultant B is correct. Wearing clean casual clothes and always having the shirt tucked in presents a professional image to the customer.

Answer C is incorrect. Only Service Consultant B is correct.

Answer D is incorrect. Service Consultant B is correct.

TASK A.1.8, B.5

16. A vehicle with 59,985 miles is in a service facility for a safety inspection prior to a pending trip. Service Consultant A says it is advisable to recommend that the timing chain be replaced due to the mileage. Service Consultant B says it is wise to recommend a 60,000-mile service be performed before the trip rather than waiting until after. Who is correct?

A. A only
B. B only
C. Both A and B
D. Neither A nor B

Answer A is incorrect. The timing chain may not require replacement at 60,000 miles. If a vehicle has a timing belt, then the service consultant needs to identify the recommended service intervals for each vehicle in order to make the correct recommendations.

Answer B is correct. Only Service Consultant B is correct. When a vehicle gets close to large maintenance services like 30,000-, 60,000-, or 90,000-mile intervals, it is advisable to make recommendations to the customer to have these services performed. Some customers like to plan for these services so it is wise to mention these services before the time arrives.

Answer C is incorrect. Only Service Consultant B is correct.

Answer D is incorrect. Service Consultant B is correct.

17. All of the following are positive characteristics of a well-run service facility EXCEPT:

TASK A.1.6, A.1.12

    A. Clean and comfortable waiting area
    B. Employees who are knowledgeable and efficient
    C. Location on a busy street
    D. Neat and organized write-up area

Answer A is incorrect. A clean and comfortable waiting area is an important aspect of a well-run service facility. Providing this service shows the customers that the business cares for the well-being of its customers.

Answer B is incorrect. Having knowledgeable and efficient employees gives the customer a sense that they are paying for professional service and will feel better about doing business there.

Answer C is correct. A good location will help increase potential business from increased exposure, but it will not guarantee that the service business will be successful. Treating customers with high levels of service and professionalism will increase the likelihood of success.

Answer D is incorrect. Having a neat and organized write-up area increases the chances of the customer having a positive experience at the service facility.

18. Service Consultant A says that obtaining accurate contact information is usually not necessary. Service Consultant B says that most customers expect the service facility to provide some form of transportation if the service takes longer than they can readily wait. Who is correct?

TASK A.1.2, A.1.5

    A. A only
    B. B only
    C. Both A and B
    D. Neither A nor B

Answer A is incorrect. It is always necessary to have accurate contact information for customers of a repair facility.

Answer B is correct. Only Service Consultant B is correct. It is wise to provide a shuttle vehicle or have loaner cars available for customers who will be without their vehicle for times longer than a couple of hours.

Answer C is incorrect. Only Service Consultant B is correct.

Answer D is incorrect. Service Consultant B is correct.

19. Which of the following would be the highest priority service on a late-model vehicle?

TASK A.2.2

    A. Hub cap replacement
    B. Steering gear replacement
    C. Cabin filter replacement
    D. Radio replacement

Answer A is incorrect. A hub cap replacement would improve the physical appearance of the vehicle, but it would not be a high-priority service procedure.

Answer B is correct. Replacing the steering gear would be a very high priority repair due to the need to properly steer the vehicle.

Answer C is incorrect. Replacing the cabin filter would improve the cleanliness of the A/C system, but it would not be considered a high priority repair.

Answer D is incorrect. Replacing the radio would improve the entertainment system, but would not be considered a high-priority repair.

TASK B.1.1

20. Which engine component regulates engine temperature?

A. Heater core
B. Thermostat
C. Water pump
D. By-pass hose

Answer A is incorrect. The heater core is the heat exchanger for the cab heating system.

Answer B is correct. The thermostat is a temperature regulating device that opens and closes to send the correct amount of coolant out of the engine and into the radiator.

Answer C is correct. The water pump is driven by either belts or chains. Its function is to move water/coolant throughout the engine area and radiator.

Answer D is incorrect. The by-pass hose provides a path for coolant to flow into the intake manifold from the water pump.

TASK B.1.2

21. Which fuel system component sprays fuel into the intake manifold when it receives a signal from the power train control module (PCM)?

A. Fuel pump
B. Fuel pressure regulator
C. Fuel injector
D. Fuel sending unit

Answer A is incorrect. The fuel pump is located in the fuel tank. It works to supply pressurized fuel up to the engine.

Answer B is incorrect. The fuel pressure regulator is the device on some fuel systems that regulates fuel pressure to the specified level depending on the fuel system requirements.

Answer C is correct. The fuel injector sprays fuel into the intake manifold when it receives an electrical signal from the PCM.

Answer D is incorrect. The fuel sending unit is a variable resistor attached to a float which is located in the fuel tank. This device sends a fuel level signal to the vehicle body electrical system which allows the fuel gauge to display fuel level.

TASK A.3.5

22. Shop quality control is a very important aspect of a successful service facility. Service Consultant A encourages her technicians to keep their work areas as clean and orderly as possible. Service Consultant B continually scans the shop area to check on the status of the vehicles. Who is practicing the correct activities?

A. A only
B. B only
C. Both A and B
D. Neither A nor B

Answer A is incorrect. Service Consultant B is also correct.

Answer B is incorrect. Service Consultant A is also correct.

Answer C is correct. Both Service Consultants are correct. A good service consultant encourages the technicians to keep the shop as clean and orderly as possible. She also continues monitoring the status of the vehicles being repaired.

Answer D is incorrect. Both Service Consultants are correct.

23. All of the following procedures would be performed during a typical 60,000-mile service EXCEPT:

   A. Replace the oil pump.
   B. Replace the spark plugs.
   C. Replace the fuel filter.
   D. Inspect and adjust the brake system.

   Answer A is correct. Replacement of the oil pump would not be included as part of a 60,000-mile service.

   Answer B is incorrect. The spark plugs are usually replaced during a 60,000-mile service.

   Answer C is incorrect. The fuel filter is typically replaced during a 60,000-mile service.

   Answer D is incorrect. The brake system is typically inspected and adjusted during a 60,000-mile service.

24. All of the following components are part of a typical starting system EXCEPT:

   A. Starter solenoid
   B. Park/neutral switch
   C. Ignition switch
   D. Voltage regulator

   Answer A is incorrect. The starter solenoid is mounted to the starter motor on most engine types. Its job is to act as a switch and to create linear movement to push the drive gear into the flywheel.

   Answer B is incorrect. The park/neutral switch is part of the starter control circuit for vehicles with automatic transmissions. This device prevents the starter from engaging while in forward or reverse gears.

   Answer C is incorrect. The ignition switch is part of the starter control circuit. This device sends the crank signal down to the starter solenoid when the driver moves the switch to the crank position.

   Answer D is correct. The voltage regulator is not part of a typical starting system. This device is a major part of the alternator/generator system. Its job is to control the charging output of the alternator/generator.

25. Which drive train component connects the transmission to the drive axle on a rear-wheel drive vehicle?

   A. Axle half shaft
   B. Driveshaft
   C. Universal joint
   D. Carrier bearing

   Answer A is incorrect. An axle half shaft is used on front-wheel drive vehicles to connect the transaxle to the drive hubs.

   Answer B is correct. The driveshaft connects the transmission to the rear-drive axle on rear-wheel drive vehicles.

   Answer C is incorrect. A universal joint is used on a driveshaft to connect it on each end. This type of component is needed to allow the drive angle to change as the rear suspension moves up and down.

   Answer D is incorrect. A carrier bearing is used on some longer wheelbase vehicles with multi-piece driveshafts.

TASK B.6.1, B.6.2

26. Service Consultant A says that items such as belts and hoses are typically considered wear items and are not typically covered by extended warranties. Service Consultant B says that major transmission components are considered wear items and are not typically covered by extended warranties. Who is correct?

    A. A only
    B. B only
    C. Both A and B
    D. Neither A nor B

    Answer A is correct. Only Service Consultant A is correct. Belts and hoses and other normal wear items are not usually covered by extended warranties.

    Answer B is incorrect. Major transmission components typically fall under the power train portion of the extended warranties. It is always advisable to call the company for approval before completing the repair.

    Answer C is incorrect. Only Service Consultant A is correct.

    Answer D is incorrect. Service Consultant A is correct.

TASK B.7.1

27. Service Consultant A says that the 10th digit of the VIN represents the year model. Service Consultant B says that the 8th digit of the VIN is the engine code. Who is correct?

    A. A only
    B. B only
    C. Both A and B
    D. Neither A nor B

    Answer A is incorrect. Service Consultant B is also correct.

    Answer B is incorrect. Service Consultant A is also correct.

    Answer C is correct. Both Service Consultants are correct. The 8th VIN digit is the engine code and the 10th VIN digit is the vehicle year.

    Answer D is incorrect. Both Service Consultants are correct.

TASK A.3.1

28. Which of the following is the most common method of communicating the customer request to the technician?

    A. Sending a text message to the technician with the customer request
    B. Announcing the information over the intercom
    C. Having the vehicle porter relay to the technician the details revealed from a follow-up phone call
    D. Writing clear and complete customer concern descriptions on the repair order

    Answer A is incorrect. Sending a text message is not a typical way of communicating a customer request to the technician.

    Answer B is incorrect. Announcing customer information over the intercom in not an advisable practice.

    Answer C is incorrect. Sending a message to the technician through a third party is not a good practice when communicating customer requests to the technician.

    Answer D is correct. Writing clear and complete instructions on the repair order is the best way to document customer concerns for the technician. This practice also creates a trail that can be retraced at a later date in the vehicle service history.

29. Which automatic transmission component acts as a fluid-coupling device between the engine and transmission?

TASK B.2.2

A. Transmission oil pump
B. Torque converter
C. Clutch pack
D. Valve body

Answer A is incorrect. The transmission oil pump creates hydraulic pressure that is used by the transmission to apply clutches and servos at the correct time to allow the transmission to automatically shift.

Answer B is correct. The torque converter is the fluid-coupling device in the automatic transmission. This component allows the engine crankshaft to rotate while the vehicle is sitting at a stop.

Answer C is incorrect. The clutch pack is one of several driving/holding hydraulic components that can be turned on and off to allow automatic shifting.

Answer D is incorrect. The valve body is the control mechanism in the automatic transmission. Most valve bodies on late-model vehicles have electronic solenoids mounted on them.

30. Service Consultant A says that some repair shops sublet radiator repairs to a specialty shop. Service Consultant B says that some repair shops sublet coolant maintenance repairs to a specialty shop. Who is correct?

TASK C.2

A. A only
B. B only
C. Both A and B
D. Neither A nor B

Answer A is correct. Only Service Consultant A is correct. It is common for most auto repair shops to sublet radiator repairs to a radiator specialty shop.

Answer B is incorrect. Coolant maintenance repairs like coolant flushes and water pump replacement is commonly done in-house by most repair facilities.

Answer C is incorrect. Only Service Consultant A is correct.

Answer D is incorrect. Service Consultant A is correct.

31. Service consultants must strive to keep an accurate customer appointment log. Service Consultant A says that it is advisable to schedule more work than the shop can complete in order to make sure the day is filled with plenty of jobs. Service Consultant B says that it is advisable to schedule only 50 percent of the shop's capacity for a day in order to make sure that all vehicles get completed on time. Who is correct?

TASK C.1, C.3

A. A only
B. B only
C. Both A and B
D. Neither A nor B

Answer A is incorrect. It is not wise to overbook the service repair shop, because this practice does not leave room for adding services to any of the repairs. Also, some of the overbooked customers will get upset because of unmet promises.

Answer B is incorrect. It is not advisable to only book at 50 percent capacity of your service shop because the productivity will drop, which will decrease the income of the shop and many of its employees.

Answer C is incorrect. Neither Service Consultant is correct.

Answer D is correct. Neither Service Consultant is correct. Accurately scheduling the right amount of work for a repair shop is a challenging and necessary skill for a service consultant. Many variables contribute to performing this skill well.

TASK A.3.7

32. Service Consultant A says that some shops require the technician to clock in on the repair order when he begins and when he finishes a repair. Service Consultant B says that some shops monitor the hours that each technician produces each week in order to track technician efficiency. Who is correct?

    A. A only
    B. B only
    C. Both A and B
    D. Neither A nor B

    Answer A is incorrect. Service Consultant B is also correct.

    Answer B is incorrect. Service Consultant A is also correct.

    Answer C is correct. Both Service Consultants are correct. It is common to require technicians to clock their time on each repair order. It is also common to monitor the productivity of each technician for planning and improvement issues.

    Answer D is incorrect. Both Service Consultants are correct.

TASK A.1.6

33. Service Consultant A says that it is wise to post the certifications and credentials of the service technicians in the write-up area. Service Consultant B says that most customers are more worried about a low cost of service than about the professionalism of the service technicians. Who is correct?

    A. A only
    B. B only
    C. Both A and B
    D. Neither A nor B

    Answer A is correct. Only Service Consultant A is correct. It provides peace of mind and a sense of value to the customers if they can see that the technicians who are performing the service on their vehicle are professional and certified.

    Answer B is incorrect. Repeat customers do put a value on the quality of the repairs that they are purchasing. These customers will not be as likely to price-shop your facility if they know they are paying a fair price for a professional technician to service their vehicle.

    Answer C is incorrect. Only Service Consultant A is correct.

    Answer D is incorrect. Service Consultant A is correct.

TASK A.2.1

34. Service Consultant A always provides vague estimates to his customers so he can quickly get the technician started on the repair. Service Consultant B provides thorough and accurate estimates to her customers so there will not be any surprises when it is time to pay the bill. Who is correct?

    A. A only
    B. B only
    C. Both A and B
    D. Neither A nor B

    Answer A is incorrect. This is a practice that will sometimes get a service consultant in a difficult situation. It is wise to be as accurate as possible with service estimates.

    Answer B is correct. Only Service Consultant B is correct. Providing accurate estimate and completion times raises customer satisfaction with a repair facility.

    Answer C is incorrect. Only Service Consultant B is correct.

    Answer D is incorrect. Service Consultant B is correct.

35. Supplemental restraint system (SRS) is another name for:

TASK B.4.2

A. Seat belts
B. Airbags
C. Upper hydraulic motor mount
D. Child safety seat anchors

Answer A is incorrect. Seat belts are primary, active restraints.

Answer B is correct. Some manufacturers also call the airbag *system supplemental inflatable restraints* (SIRs). Many cars for years had only primary restraints; supplemental restraints increase safety and cannot be used as a substitute for primary restraints. US legislation passed in 1989 mandated that passenger cars be equipped with passive restraints, or restraints that do not require action by the driver (e.g., using a seat belt). Both airbags and automatic seatbelts are considered passive restraints and thus one or the other is commonly included on late-model vehicles.

Answer C is incorrect. This component controls the top of the engine.

Answer D is incorrect. These devices are part of the primary restraint system.

36. Service Consultant A says that an example of a feature of an oil change is the brand of oil used. Service Consultant B says that an example of a benefit of an oil change is longer engine life. Who is right?

TASK A.2.5

A. A only
B. B only
C. Both A and B
D. Neither A nor B

Answer A is incorrect. Service Consultant B is also correct.

Answer B is incorrect. Service Consultant A is also correct.

Answer C is correct. Both Service Consultants are correct. This is an example of the difference between features and benefits. The *feature* of a service describes what processes take place during that service. The *benefits* of a service are the results of having the service performed.

Answer D is incorrect. Both Service Consultants are correct.

37. Service Consultant A says that adding a description of the work performed adds value to the repair. Service Consultant B says that the VIN number may be used to find part applications. Who is right?

TASK A.1.14, B.7.1

A. A only
B. B only
C. Both A and B
D. Neither A nor B

Answer A is incorrect. Service Consultant B is also correct.

Answer B is incorrect. Service Consultant A is also correct.

Answer C is correct. Both Service Consultants are correct. Customers appreciate a simple explanation of the work that is performed on their vehicle. Using the VIN number is an effective way to assist the parts professionals in finding the correct parts for the vehicle.

Answer D is incorrect. Both Service Consultants are correct.

TASK B.5.2

38. Service Consultant A says that the oil change reminder light should be reset by the technician when the engine oil is changed. Service Consultant B says that the oil life percentage reminder system should be reset when the oil is changed. Who is correct?

   A. A only
   B. B only
   C. Both A and B
   D. Neither A nor B

Answer A is incorrect. Service Consultant B is also correct.

Answer B is incorrect. Service Consultant A is also correct.

Answer C is correct. Both Service Consultants are correct. Oil change reminder lights and oil life percentage reminders should both be reset at the time of the oil change. Customers expect that these types of systems be reset at the time that the oil change occurs. Most customers will not know how to reset these systems and they expect the repair shop to provide this as part of the oil change.

Answer D is incorrect. Both Service Consultants are correct.

TASK B.5.1

39. All of the following are elements of an oil change service EXCEPT:

   A. Test the pH level of the coolant.
   B. Drain the oil and refill with the correct type and amount of new oil.
   C. Remove the oil filter and replace with the correct new oil filter.
   D. Check all engine fluid levels and notify the owner of the systems that were low.

Answer A is correct. It is not a normal practice to test the pH level of the coolant during a routine oil change service. This test would likely be a part of a cooling system service.

Answer B is incorrect. Draining and refilling the engine oil is a normal practice that occurs during an oil change service.

Answer C is incorrect. Removing and replacing the oil filter is a normal practice that occurs during an oil change service.

Answer D is incorrect. Checking all fluid levels is a normal practice that happens during an oil change service.

TASK B.3.2

40. Which hydraulic brake component converts linear movement into hydraulic pressure that is sent to the wheels to stop the vehicle?

   A. Metering valve
   B. Proportioning valve
   C. Wheel cylinder
   D. Master cylinder

Answer A is incorrect. The metering valve is a hold-off valve that blocks pressure to the front wheels until the rear wheel cylinders begin to apply the rear brake shoes.

Answer B is incorrect. The proportioning valve is a device that limits pressure to the rear wheels during heavy braking in order to help prevent rear wheel lockup.

Answer C is incorrect. The wheel cylinder is a hydraulic device used in drum brake systems that pushes the brake shoes into the brake drum when the brake pedal is depressed.

Answer D is correct. The master cylinder mounts to the firewall and mechanically connects to the brake pedal. As the brake pedal moves downward, the linear movement is applied to the pistons in the master cylinder, which creates hydraulic pressure that gets pumped down to the wheels to stop the vehicle.

41. Service Consultant A says that camber is the alignment angle that measures the inward or outward tilt of the tire. Service Consultant B says that caster is the alignment angle that measures the forward or rearward tilt of the steering axis. Who is correct?

TASK B.3.2

A. A only
B. B only
C. Both A and B
D. Neither A nor B

Answer A is incorrect. Service Consultant B is also correct.

Answer B is incorrect. Service Consultant A is also correct.

Answer C is correct. Both Service Consultants are correct. Camber is the alignment angle that measures the inward or outward tilt of the tire. Positive camber means that the tire is leaning out at the top and negative camber means that the tire is leaning in at the top. Caster is the alignment angle that measures the forward or rearward tilt of the steering axis. Positive caster means that the steering axis is leaning rearward and negative caster means that the steering axis is leaning forward.

Answer D is incorrect. Both Service Consultants are correct.

42. Which of the following items would be the LEAST LIKELY component to be replaced during a clutch replacement?

TASK B.2.3

A. Throw out bearing
B. Pressure plate
C. Input gear
D. Clutch disc

Answer A is incorrect. The throw out bearing is a typical component that is replaced during a clutch replacement.

Answer B is incorrect. The pressure plate is a typical component that is replaced during a clutch replacement.

Answer C is correct. The input gear is located inside the transmission and would not normally be replaced during a clutch replacement.

Answer D is incorrect. The clutch disc is a typical component that is replaced during a clutch replacement.

43. Which of the ignition system components creates the high-voltage spike that creates the energy to fire the spark plug?

TASK B.1.2

A. Ignition module
B. Coil pack
C. Crank sensor
D. Plug wire

Answer A is incorrect. The ignition module is an electronic device that sends a signal to the coil pack that causes it to create voltage at the correct time.

Answer B is correct. The coil pack is the device that creates the high-voltage energy that is needed to fire the spark plug. The crank sensor, engine control module (ECM), and the ignition module operate together to signal the coil to fire at the correct time.

Answer C is incorrect. The crank sensor is located in the engine block that creates a voltage signal that is proportional to engine RPM. This signal is typically sent to the ECM, which then sends a signal to the ignition module.

Answer D is incorrect. The plug wire connects the coil pack electrically to the spark plug.

TASK B.1.3

44. Service Consultant A says that the electric fuel pump is located inside the fuel tank. Service Consultant B says that the fuel filter is located under the intake manifold plenum. Who is correct?

    A. A only
    B. B only
    C. Both A and B
    D. Neither A nor B

Answer A is correct. Only Service Consultant A is correct. On vehicles with electronic fuel injection, the electric fuel pump is located inside the fuel tank. Some vehicles have an access panel that can be removed from the floor pan of the vehicle to service the pump. Some vehicles require the fuel tank to be removed to service the fuel pump.

Answer B is incorrect. The fuel filter would not typically be located under the intake manifold plenum. The filter is located somewhere in the fuel line leading up to the engine area.

Answer C is incorrect. Only Service Consultant A is correct.

Answer D is incorrect. Service Consultant A is correct.

TASK A.3.6

45. Service Consultant A carefully documents the services that are performed on each repair order prior to notifying the customer that the vehicle is done. Service Consultant B documents the recommended services on the repair order in order to communicate this information to the customer. Who is correct?

    A. A only
    B. B only
    C. Both A and B
    D. Neither A nor B

Answer A is incorrect. Service Consultant B is also correct.

Answer B is incorrect. Service Consultant A is also correct.

Answer C is correct. Both Service Consultants are correct. The services that are performed, as well as the recommended services, should be written on the repair order so the customer has a record of what was done as well as what the shop recommends in the future.

Answer D is incorrect. Both Service Consultants are correct.

TASK A.1.8, A.2.4

46. All of the following examples would be considered a benefit of recommending additional services to a current customer who has left his/her car at your shop EXCEPT:

    A. The shop can stay open late to finish the repair.
    B. The shop is more profitable from the increased sales of parts.
    C. The shop is more profitable from the increased sales of labor.
    D. The customer will begin to trust that the shop is looking out for his/her well-being and safety.

Answer A is correct. Staying open late to finish a repair is not a benefit of recommending additional services. Asking the necessary employees to stay late on a regular basis would not be good business practice due to paying overtime to these employees.

Answer B is incorrect. The shop will be more profitable if some additional parts are sold to current customers during their visit.

Answer C is incorrect. The shop will be more profitable if some additional labor fees are sold to current customers during their visit.

Answer D is incorrect. Customer trust can be built by making recommendations of needed additional services. Service consultants must be sure to only recommend what is needed based upon proven signs of wear or maintenance schedules.

47. A customer is objecting to having additional needed maintenance service performed. Service Consultant A explains that the vehicle will likely break down very soon and strand the customer in a dangerous location. Service Consultant B explains the value of having a well-maintained vehicle. Who is correct?

TASK A.2.4, A.2.6

A. A only
B. B only
C. Both A and B
D. Neither A nor B

Answer A is incorrect. Scare tactics should not be regularly used as a method of motivating customers to authorize additional needed maintenance.

Answer B is correct. Only Service Consultant B is correct. Customers should be dealt with in a positive manner. This can be done by explaining the benefits of having a well-maintained vehicle.

Answer C is incorrect. Only Service Consultant B is correct.

Answer D is incorrect. Service Consultant B is correct.

48. A customer is at the repair shop describing the problems with his/her vehicle. Service Consultant A asks detailed questions of the customer to determine what the main concern is. Service Consultant B asks questions about the time, temperature, and the frequency with which the problem is occurring. Who is correct?

TASK A.1.3, A.2.3

A. A only
B. B only
C. Both A and B
D. Neither A nor B

Answer A is incorrect. Service Consultant B is also correct.

Answer B is incorrect. Service Consultant A is also correct.

Answer C is correct. Both Service Consultants are correct. Good questions asked by the service consultant will increase the accuracy of the repair order. Details about the problem including the time, the temperature as well as the frequency will assist the technician in diagnosing the root cause of the problem.

Answer D is incorrect. Both Service Consultants are correct.

49. Which of the following would be the most critical safety service to be performed on a late-model vehicle?

TASK A.2.2

A. Oil change
B. Door panel replacement
C. Brake pad replacement
D. Air conditioner service

Answer A is incorrect. Oil changes are an important service to have performed. However, there is typically a range of time and mileage that oil changes can be performed at.

Answer B is incorrect. A door panel replacement would not be considered a critical repair service.

Answer C is correct. The brake system is a very critical system that must be maintained in order for the vehicle to be safe. This repair should be performed when the pads are worn out.

Answer D is incorrect. The air conditioner is a customer comfort system and would not be considered a critical safety service.

TASK A.2.5

50. All of the following would be benefits of having service repair work done at the same quality repair shop every time EXCEPT:

    A. The oil changes are cheaper.
    B. A service history is developed for the vehicle.
    C. A trustful relationship is developed between the owner and the repair shop's employees.
    D. The technicians become familiar with the vehicle.

Answer A is correct. Basic quick services may not always be cheaper at a given repair location, but if the customer gets a fair price for dependable and honest repair services, then he will likely become a repeat customer.

Answer B is incorrect. Building a repair history at a repair shop is a benefit to the customer, because the service consultant and the technician can access these records during the visit to help in the diagnosis and maintenance schedules on the vehicle.

Answer C is incorrect. Building a relationship with the repair facility is a benefit of returning to the same place for vehicle repair each time it is needed.

Answer D is incorrect. There is a benefit to the service technician being familiar with a vehicle. She will be more likely to provide high levels of expertise and good service on a familiar vehicle.

PREPARATION EXAM 2—ANSWER KEY

1. D
2. C
3. D
4. C
5. C
6. B
7. B
8. B
9. B
10. C
11. C
12. A
13. B
14. B
15. A
16. B
17. D
18. B
19. D
20. A
21. A
22. C
23. C
24. D
25. B
26. B
27. C
28. C
29. A
30. B
31. B
32. A
33. D
34. D
35. B
36. A
37. B
38. D
39. D
40. A
41. B
42. C
43. D
44. C
45. B
46. B
47. A
48. C
49. B
50. A

PREPARATION EXAM 2—EXPLANATIONS

1. Service Consultant A says that a ringing phone always takes precedent over a customer standing in front of him. Service Consultant B says that if a customer walks in while you are on the phone, you should quickly take a message and deal with the customer in front of you. Who is correct?

TASK A.1.1

A. A only

B. B only

C. Both A and B

D. Neither A nor B

Answer A is incorrect. If a service consultant is directly dealing with a customer in his/her presence, then all of their attention should be given to the customer. It is wise to have other counter workers available to answer the phone when the service consultant is working with customers onsite. At the very least, all service facilities should have a voice messaging system that allows the calling customer to leave a message.

Answer B is incorrect. If a service consultant is having a conversation with a customer on the phone and another customer walks up, the consultant should acknowledge their presence by a nod or wave and then complete the phone conversation.

Answer C is incorrect. Neither Service Consultant is correct.

Answer D is correct. Neither Service Consultant is correct. Phone customers and walk-in customers are both very important to the vitality of a service facility. A service consultant should do everything possible to handle all customers with professionalism and integrity.

TASK A.1.2

2. All of the following vehicle information items should be collected when preparing a vehicle repair order EXCEPT:

   A. VIN
   B. Color
   C. Transmission model
   D. Vehicle model

Answer A is incorrect. The VIN number is a very important piece of information to include on the vehicle repair order. The VIN has 17 digits that give information such as the manufacturer, the year, the engine, and the body style, as well as the sequence in which the vehicle was built.

Answer B is incorrect. The vehicle color should be included on the repair order to limit confusion while the vehicle is at the repair facility.

Answer C is correct. The transmission model is important in some applications, but it is not a typical piece of information that would appear on every repair order.

Answer D is incorrect. The vehicle model should be present on the repair order to help the staff identify the vehicle while it is at the repair facility.

TASK B.7.1

3. Which of the following items is NOT listed as a digit on the VIN?

   A. Country of origin
   B. Body style
   C. Vehicle year
   D. Vehicle build month

Answer A is incorrect. The country of origin is listed as the first digit of the VIN.

Answer B is incorrect. The vehicle body is listed as the 6th digit of the VIN.

Answer C is incorrect. The vehicle year is listed as the 10th digit of the VIN.

Answer D is correct. The month when the vehicle was built is not listed in the VIN. This information can be found on the B-pillar on the driver's side door area.

TASK A.3.3, A.3.4, C.1, C.3

4. A vehicle has been dropped off at the repair shop with instructions from the owner to diagnose a brake noise and then call with an estimate and completion time. Service Consultant A checks the appointment log and the status of other jobs with the technician before predicting a completion time. Service Consultant B verifies the price and availability of the parts before predicting a completion time. Who is correct?

   A. A only
   B. B only
   C. Both A and B
   D. Neither A nor B

Answer A is incorrect. Service Consultant B is also correct.

Answer B is incorrect. Service Consultant A is also correct.

Answer C is correct. Both Service Consultants are correct. It is advisable to check on the flow of work in the shop prior to predicting a completion time. It is also wise to check on the price and availability of the parts prior to predicting a completion time of the repair.

Answer D is incorrect. Both Service Consultants are correct.

5. Which engine component will benefit the most from regular oil and filter changes?

TASK B.1.1

   A. Thermostat
   B. Cylinder head
   C. Crankshaft
   D. Piston pin

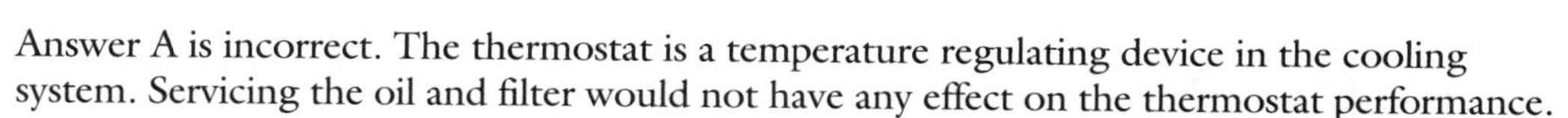

Answer A is incorrect. The thermostat is a temperature regulating device in the cooling system. Servicing the oil and filter would not have any effect on the thermostat performance.

Answer B is incorrect. The cylinder head is a major engine component that makes up the upper part of the combustion chamber. The cylinder head houses the intake and exhaust valves. Servicing the oil and filter would not have much benefit directly for the cylinder head.

Answer C is correct. The crankshaft would benefit the most from regular oil and filter changes. A thin layer of oil is the only thing between the crankshaft and the bearings. If the oil gets dirtier than normal from not changing the oil, the crankshaft and its bearings will begin to wear prematurely.

Answer D is incorrect. The piston pin connects the piston to the connecting rod. Servicing the oil and filter would not benefit the performance of the piston pin.

6. All of the following are positive reasons a service consultant could use to finalize a service sale EXCEPT:

TASK A.2.6

   A. The vehicle will be more reliable if the correct service and maintenance are performed on time.
   B. The owner will be more likely to trade the vehicle because he is dissatisfied with it.
   C. The vehicle will hold its value better if good service practices are followed.
   D. The owner will be safer driving a well-maintained vehicle.

Answer A is incorrect. A well-maintained vehicle will be more reliable.

Answer B is correct. This is not a positive reason a service consultant might use when finalizing a service sale.

Answer C is incorrect. A well-maintained vehicle will hold its value better.

Answer D is incorrect. A well-maintained vehicle is safer and less likely to break down without warning.

7. Service Consultant A says that some front-wheel drive vehicles have a full steel frame that is similar to rear-wheel drive vehicles. Service Consultant B says that some front-wheel drive vehicles use an electrically assisted power steering system. Who is correct?

TASK B.3.2

   A. A only
   B. B only
   C. Both A and B
   D. Neither A nor B

Answer A is incorrect. Front-wheel drive vehicles do not typically have full steel frames. Instead, these vehicles use a cradle style frame that supports the engine and transmission and mounts to the vehicle body.

Answer B is correct. Only Service Consultant B is correct. Some late-model vehicles use an electrically assisted power steering system. These vehicles will not have a hydraulic power steering system.

Answer C is incorrect. Only Service Consultant B is correct.

Answer D is incorrect. Service Consultant B is correct.

TASK A.3.3

8. Service Consultant A does not check for the availability of the repair parts before notifying the customer. Service Consultant B does not call the customer until after checking the availability and price of the repair parts. Who is correct?

   A. A only
   B. B only
   C. Both A and B
   D. Neither A nor B

Answer A is incorrect. Notifying the customer before checking on the availability of the repair parts is not advisable. The customer will need to be given a completion time if he/she agrees to have the repair completed.

Answer B is correct. Only Service Consultant B is correct. It is advisable to check the price and availability of the repair parts prior to calling the customer.

Answer C is incorrect. Only Service Consultant B is correct.

Answer D is incorrect. Service Consultant B is correct.

TASK B.4.2

9. Service Consultant A says that the power windows operate by using bi-directional solenoids to move the window up and down. Service Consultant B says that the power seats use bi-directional permanent magnet (PM) motors to move the seat to the desired location. Who is correct?

   A. A only
   B. B only
   C. Both A and B
   D. Neither A nor B

Answer A is incorrect. Solenoids do not have enough range to be used in power windows.

Answer B is correct. Only Service Consultant B is correct. The power seats use bi-directional PM motors that change directions by reversing the polarity through the motor.

Answer C is incorrect. Only Service Consultant B is correct.

Answer D is incorrect. Service Consultant B is correct.

TASK A.1.12, A.1.13

10. All of the following questions should be asked on a customer follow-up call EXCEPT:

   A. Were you pleased with the way you were greeted at the repair shop?
   B. Were you pleased with the neatness of the waiting area at the repair shop?
   C. Were you pleased with the brand of coffee that was served in the waiting room?
   D. Were you pleased with the quality of the repair that was performed by the facility?

Answer A is incorrect. It is a good idea to ask the customer about how they were greeted when performing a follow-up call.

Answer B is incorrect. It is a good idea to ask the customer about the neatness of the waiting areas and the shop when performing a follow-up call.

Answer C is correct. The customer's preference with respect to free coffee is not a typical piece of information that would be retrieved on a follow-up call.

Answer D is incorrect. It is a good idea to ask the customer about the perceived quality of the repair when performing a follow-up call.

11. Service Consultant A says that the vehicle battery cable should never be disconnected while the engine is running. Service Consultant B says the vehicle battery can run down if the vehicle is stored for long periods of time. Who is correct?

TASK B.1.3

A. A only

B. B only

C. Both A and B

D. Neither A nor B

Answer A is incorrect. Service Consultant B is also correct.

Answer B is incorrect. Service Consultant A is also correct.

Answer C is correct. Both Service Consultants are correct. Disconnecting the battery cable with the engine running can potentially cause damage to the computer systems on the car. The battery should be disconnected to prevent drain if the vehicle is going to be stored for long periods of time.

Answer D is incorrect. Both Service Consultants are correct.

12. All of the following items should be added to the estimate when calculating the total cost for a repair EXCEPT:

TASK A.2.1

A. Labor amount paid to the technician

B. Parts total

C. Labor total

D. Sublet repair costs

Answer A is correct. The labor amount that the technician is paid is not calculated when figuring the amount of an estimate.

Answer B is incorrect. The total charges for the repair parts would be included on a repair estimate.

Answer C is incorrect. The total labor charges would be included on the repair estimate.

Answer D is incorrect. The total sublet repair costs would be included on the repair estimate.

13. A customer is requesting information about the warranty of the work that is performed by a repair shop. Service Consultant A says that some remanufactured engines carry a warranty for as long as 3 months or 3,000 miles. Service Consultant B says that some batteries carry a three-year free replacement warranty. Who is correct?

TASK B.6.1

A. A only

B. B only

C. Both A and B

D. Neither A nor B

Answer A is incorrect. Remanufactured engines will typically have longer warranties than 3 months or 3,000 miles. Warranties will vary greatly among components and rebuilders. It would be advisable to develop a routine of using the same company for remanufactured engines in order to be very familiar with the details of the warranties and costs.

Answer B is correct. Only Service Consultant B is correct. Some premium batteries come with a three-year free replacement warranty.

Answer C is incorrect. Only Service Consultant B is correct.

Answer D is incorrect. Service Consultant B is correct.

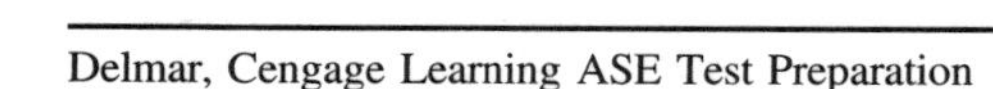

TASK A.2.3

14. Service Consultant A is offended if a customer asks detailed questions about why a particular service is being recommended. Service Consultant B provides clear and understandable answers to detailed questions. Who is correct?

    A. A only
    B. B only
    C. Both A and B
    D. Neither A nor B

Answer A is incorrect. A service consultant should be glad to answer detailed questions about any service that he is recommending. If the question is too technical, then the service consultant can invite the technician to join the conversation.

Answer B is correct. Only Service Consultant B is correct. The service consultant should strive to answer any questions that a customer might have in a manner that the customer feels comfortable.

Answer C is incorrect. Only Service Consultant B is correct.

Answer D is incorrect. Service Consultant B is correct.

TASK B.2.3

15. What is the LEAST LIKELY step that would be performed during an automatic transmission fluid and filter service?

    A. Install new friction discs into the clutch packs.
    B. Remove the transmission oil pan and drain the old fluid into a suitable container.
    C. Replace or clean the old transmission filter.
    D. Install the transmission pan with a new gasket.

Answer A is correct. New friction discs would be installed during a complete transmission overhaul procedure.

Answer B is incorrect. The transmission pan is removed while the fluid is drained into a suitable container during a transmission service.

Answer C is incorrect. It is common to replace or clean the old transmission filter during this service.

Answer D is incorrect. A new gasket should be used when installing the transmission pan. The bolts should be tightened to the correct torque value.

TASK A.1.7

16. Service Consultant A says that it is not necessary to know what has been done to the vehicle in the past. Service Consultant B says that some shops have computerized software that stores the service data on each vehicle that has been in the repair facility. Who is correct?

    A. A only
    B. B only
    C. Both A and B
    D. Neither A nor B

Answer A is incorrect. Ignoring or being unaware of the service history on any vehicle puts the repair facility at a big disadvantage. Knowing what has been done in the past helps the service consultant and the technician evaluate what current services are needed.

Answer B is correct. Only Service Consultant B is correct. Computer shop software has been used for many years as a way to generate repair orders as well as store the service history for the facility.

Answer C is incorrect. Only Service Consultant B is correct.

Answer D is incorrect. Service Consultant B is correct.

17. All of the following items require a reset procedure by a repair shop after services are performed EXCEPT:

TASK B.5.2

A. Oil change reminder light
B. Oil life percentage reminder
C. Check engine light
D. Low coolant level indicator

Answer A is incorrect. Some vehicles have an oil change reminder light that should be reset at each oil change.

Answer B is incorrect. Some vehicles have an oil life percentage reminder that should be reset at each oil change.

Answer C is incorrect. The diagnostics codes should be cleared after a repair has been made. This will turn off the check engine light.

Answer D is correct. The low coolant level indicator only comes on when the coolant level is low and will typically go out when the problem is corrected.

18. Which of the following repair procedures would be the LEAST LIKELY to be considered a high priority repair?

TASK A.2.2

A. Replacement of a worn tie rod end
B. Replacement of the rear wiper blade
C. Replacement of a tire with the steel showing
D. Front brake pads replaced

Answer A is incorrect. A worn tie rod end is a high-priority repair since it could come loose and cause the driver to lose control of the vehicle.

Answer B is correct. A worn rear wiper blade would not cause the vehicle to be dangerous to drive.

Answer C is incorrect. A severely worn tire would be a great danger to continue using. This repair would need to be made immediately.

Answer D is incorrect. Worn brake pads would make the vehicle dangerous to drive, so they would need to be replaced very soon.

19. Service Consultant A says that a manufacturer recall/campaign is a technical document created to help technicians to repair pattern failures more quickly. Service Consultant B says that a technical service bulletin (TSB) is a program that manufacturers create to invite customers to bring their vehicles back in for a free repair in order to correct a safety fault in the vehicle. Who is correct?

TASK B.6.3

A. A only
B. B only
C. Both A and B
D. Neither A nor B

Answer A is incorrect. A manufacturer recall/campaign is a program that manufacturers create to invite customers to bring their vehicles back in for a free repair in order to correct a safety fault in the vehicle.

Answer B is incorrect. A TSB is a technical document created to help technicians repair pattern failures more quickly.

Answer C is incorrect. Neither Service Consultant is correct.

Answer D is correct. Neither Service Consultant is correct. A good service consultant needs to be familiar with the technical terms used daily in the service repair business. A manufacturer recall/campaign is a program that manufacturers create to invite customers to bring their vehicles back in for a free repair in order to correct a safety fault in the vehicle. A technical service bulletin is a technical document created to help technicians repair pattern failures more quickly.

TASK A.1.6, A.2.5

20. A first-time customer walks into a repair shop and asks the service consultants at the desk to explain the benefits and features of having repair work done at that shop. Service Consultant A says that one benefit of the shop is that all of the technicians are ASE certified in the areas in which they perform repair work. Service Consultant B says that the employees at the shop are more honest than at other shops in town. Who is correct?

A. A only
B. B only
C. Both A and B
D. Neither A nor B

Answer A is correct. Only Service Consultant A is correct. Having ASE certified technicians is a worthwhile benefit to mention to a customer.

Answer B is incorrect. It is not recommended to talk negatively about other shops in order to make your own shop look better. It is better to just mention the positive traits of your shop when explaining the benefits of using your facility.

Answer C is incorrect. Only Service Consultant A is correct.

Answer D is incorrect. Service Consultant A is correct.

TASK C.3

21. All of the following items should be entered into the customer appointment log EXCEPT:

A. Tire size
B. Vehicle year, make, and model
C. Estimated time of repair
D. Customer name

Answer A is correct. The tire size of each vehicle is not a necessary piece of information for the customer appointment log.

Answer B is incorrect. The vehicle year, make, and model are usually present on the customer appointment log.

Answer C is incorrect. A rough time of repair estimate would be on the customer appointment log. This time will usually not be extremely accurate, but it helps to plan the number of vehicles that need to be planned for each day.

Answer D is incorrect. The customer name would always be on the appointment log.

TASK A.3.2

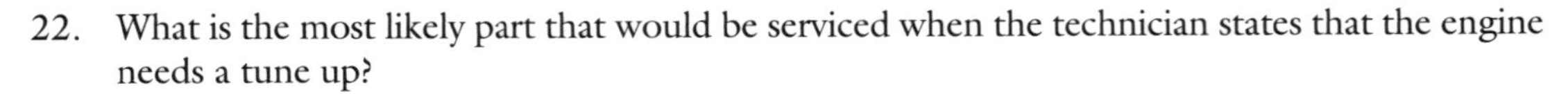

22. What is the most likely part that would be serviced when the technician states that the engine needs a tune up?

A. Piston rings
B. Valve cover gaskets
C. Fuel filter
D. Oxygen sensor

Answer A is incorrect. The piston rings would only be replaced during a major engine rebuild procedure.

Answer B is incorrect. The valve cover gaskets would be replaced if they were leaking oil.

Answer C is correct. The fuel filter is a normal component that is replaced during an engine tune-up. In addition, the technician would typically replace the air filter and spark plugs.

Answer D is incorrect. The oxygen sensor would only be replaced if there were a problem with the operation of the sensor. It would not normally be replaced during an engine tune up.

23. Service Consultant A verifies whether a customer is a first-time customer when scheduling an appointment. Service Consultant B verifies whether a customer is a potential warranty customer when scheduling an appointment. Who is correct?

TASK A.1.11

A. A only
B. B only
C. Both A and B
D. Neither A nor B

Answer A is incorrect. Service Consultant B is also correct.

Answer B is incorrect. Service Consultant A is also correct.

Answer C is correct. Both Service Consultants are correct. It is advisable to verify the type of customer as the appointment is made in order to better serve each customer at the highest level.

Answer D is incorrect. Both Service Consultants are correct.

24. All of the following are appropriate greeting practices for a service consultant EXCEPT:

TASK A.1.4

A. Smile
B. Cordial attitude
C. Handshake
D. Fist bump

Answer A is incorrect. Greeting the customer with a smile helps the customer feel more at ease.

Answer B is incorrect. Having a cordial attitude allows the customer to feel welcome when being greeted by the service consultant.

Answer C is incorrect. A handshake or some other type of physical welcome gesture shows the customer that the service consultant is glad to be able to offer service to him/her.

Answer D is correct. A fist bump is not an appropriate greeting characteristic for a professional service consultant.

25. Service Consultant A feels that she is not receiving the support from the manager that she needs, so she is usually negative with the technicians and customers. Service Consultant B has several items that concern him about the support he is receiving from the manager, so he schedules time for a face-to-face meeting with the manager. Who is correct?

TASK A.1.12, A.3.8

A. A only
B. B only
C. Both A and B
D. Neither A nor B

Answer A is incorrect. Expressing negative feelings to technicians and customers is not a good practice for a service consultant. It is wise to schedule a meeting with the manager or supervisor to discuss the causes for any negative feelings.

Answer B is correct. Only Service Consultant B is correct. It is good to keep the lines of communication open as a service consultant. Performing this job involves high levels of stress, so it is important to continually communicate any problems or issues to the owner or supervisor.

Answer C is incorrect. Only Service Consultant B is correct.

Answer D is incorrect. Service Consultant B is correct.

TASK B.1.1

26. A vehicle is in the repair shop with a check engine light that stays illuminated. The technician has checked the computer system for a problem and diagnosed the problem to be a faulty oxygen sensor. Service Consultant A says that oxygen sensors measure the amount of oxygen in the air as it passes through the intake manifold. Service Consultant B says that the oxygen sensor can deteriorate over time and requires periodic replacement. Who is correct?

   A. A only
   B. B only
   C. Both A and B
   D. Neither A nor B

Answer A is incorrect. The oxygen sensor measures the amount of oxygen in the exhaust after the combustion cycle.

Answer B is correct. Only Service Consultant B is correct. Oxygen sensors can deteriorate as they age due to their location in the exhaust system.

Answer C is incorrect. Only Service Consultant B is correct.

Answer D is incorrect. Service Consultant B is correct.

TASK B.1.2

27. From which source does the vehicle electrical system receive its energy?

   A. Starter
   B. Engine control module (ECM)
   C. Alternator
   D. Body control module (BCM)

Answer A is incorrect. The starter cranks the engine over when the key/ignition switch is moved to start.

Answer B is incorrect. The ECM monitors inputs from various sensors on the engine to control engine functions.

Answer C is correct. The alternator operates to provide the electrical power while the vehicle is running. The alternator also charges the vehicle battery while running. During times of low engine speed and high electrical loads, the vehicle battery helps supply the electrical current that is not produced by the alternator.

Answer D is incorrect. The BCM monitors inputs from all over the body to control body functions.

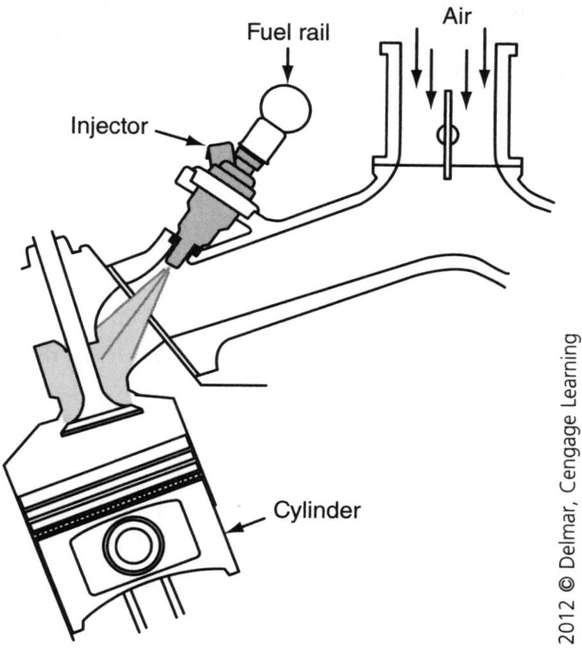

28. All of the following statements are correct about the figure above EXCEPT:

TASK B.1.2

A. The electric fuel pump supplies pressurized fuel to the fuel injector.
B. The throttle plate opens and closes as the accelerator pedal is depressed.
C. The injector supplies fuel for all of the cylinders on the engine.
D. The injector sprays fuel, which mixes with the air before entering the combustion chamber.

Answer A is incorrect. The electric fuel pump is mounted in the fuel tank and it supplies pressurized fuel to the fuel injector.

Answer B is incorrect. The throttle plate is opened and closed as the accelerator pedal is depressed. Some vehicles use a cable to accomplish this action, while many late-model vehicles use an electronic motor that is controlled by the power train control module (PCM).

Answer C is correct. The injector in the picture is a port fuel injector, which means it only provides fuel for one cylinder. There will be an injector for each cylinder in this type of fuel system.

Answer D is incorrect. The fuel injector sprays fuel into the intake manifold where it mixes with the air. The intake valve opens to allow this charged mixture into the combustion chamber.

29. A customer visits a repair facility with a transmission problem. The transmission is not shifting correctly, and the repair will likely take several hours. Service Consultant A offers to let the customer use the phone to call someone to pick him/her up. Service Consultant B directs the customer to the waiting room. Who is correct?

TASK A.1.5

A. A only
B. B only
C. Both A and B
D. Neither A nor B

Answer A is correct. Only Service Consultant A is correct. Allowing the customer to call someone to pick him/her up is sometimes an appropriate form of handling their transportation problems.

Answer B is incorrect. It is not a good practice to give unreasonable expectations to a customer by having him/her wait on-site. If a repair is going to take several hours, then alternate transportation needs to be arranged for the customer.

Answer C is incorrect. Only Service Consultant A is correct.

Answer D is incorrect. Service Consultant A is correct.

TASK A.1.14

30. A customer arrives to pick her vehicle after extensive repairs are made. Service Consultant A directs the customer to the cashier to pay her bill and has the vehicle porter pull the vehicle around to the front of the building. Service Consultant B carefully explains all of the repairs and charges to the customer before directing her to the cashier to pay the bill. Who is correct?

A. A only
B. B only
C. Both A and B
D. Neither A nor B

Answer A is incorrect. The service consultant should explain the services and charges and seek questions from the customer before directing her to pay her bill.

Answer B is correct. Only Service Consultant B is correct. All interactions with the customer must be professional and thorough in order for the customer to feel good about doing business with a shop. An explanation of the repairs and charges is a good way to finish this business transaction.

Answer C is incorrect. Only Service Consultant B is correct.

Answer D is incorrect. Service Consultant B is correct.

TASK B.2.2

31. Service Consultant A says that the engine flywheel drives the clutch disc when the clutch pedal is pressed. Service Consultant B says that the pressure plate is the part of the system that compresses the clutch disc into the flywheel when the clutch pedal is not pressed. Who is correct?

A. A only
B. B only
C. Both A and B
D. Neither A nor B

Answer A is incorrect. When the clutch pedal is pressed, the pressure plate releases the pressure, which disconnects any power from being transferred.

Answer B is correct. Only Service Consultant B is correct. The pressure plate is the device that adds pressure to the clutch disc to force it against the flywheel. This causes power to be transmitted from the engine into the manual transmission when the clutch pedal is released.

Answer C is incorrect. Only Service Consultant B is correct.

Answer D is incorrect. Service Consultant B is correct.

TASK A.2.4

32. Each of the following is a benefit of recommending additional services to a current customer who has left his/her car at your shop EXCEPT:

A. The customer will need a second opinion.
B. The shop is more profitable from the increased sales of parts.
C. The shop is more profitable from the increased sales of labor.
D. The customer will begin to trust that the shop is looking out for his or her well-being and safety.

Answer A is correct. Once the relationship is built with the customer, he/she will trust the repair facility and will not likely request a second opinion.

Answer B is incorrect. The shop will be more profitable if some additional parts are sold to current customers during their visit.

Answer C is incorrect. The shop will be more profitable if some additional labor fees are sold to current customers during their visit.

Answer D is incorrect. Customer trust can be built by making recommendations for needed additional services. Service consultants must be sure to only recommend what is needed based upon proven signs of wear or maintenance schedules.

33. Service Consultant A says that all the automatic transmission fluid should be changed once each year or every 12,000 miles. Service Consultant B says that all transmission fluid is designed to last for the life of the vehicle if the vehicle is not used to pull a heavy load. Who is correct?

TASK A.1.8, B.2.2

A. A only
B. B only
C. Both A and B
D. Neither A nor B

Answer A is incorrect. The fluid service intervals will vary depending on the vehicle type as well as how the customer uses the vehicle.

Answer B is incorrect. Transmission fluid needs to be changed at maintenance intervals depending on how the customer uses the vehicle.

Answer C is incorrect. Neither Service Consultant is correct.

Answer D is correct. Neither Service Consultant is correct. Each vehicle and customer should be advised according to vehicle type and driving habits.

34. Which of the following brake system components converts linear movement into hydraulic pressure that is sent to the wheel brakes?

TASK B.3.2

A. Wheel cylinder
B. Vacuum booster
C. Caliper
D. Master cylinder

Answer A is incorrect. The wheel cylinder is used on drum brake systems and is mounted at the backing plate. This component receives hydraulic pressure from the master cylinder to push the brake shoes against the brake drums.

Answer B is incorrect. The vacuum booster is part of the brake assist (power brakes) system. It mounts at the firewall right behind the master cylinder.

Answer C is incorrect. The caliper is used on disc brake systems to clamp the brake pads down on the rotor when hydraulic pressure is received from the master cylinder.

Answer D is correct. The master cylinder receives linear movement from the brake pedal, which moves the pistons to create hydraulic pressure that is routed to the wheel cylinders/calipers.

35. What are the most likely steps that would be followed by a service consultant to verify the accuracy of a repair order prior to calling the customer for repair authorization?

TASK A.1.10

A. Add the labor and the parts on the back of the repair order.
B. Input all the repair-related data into the computer and verify the availability of the parts.
C. Add the cost of the parts, the cost of the labor, and the cost of sublet repairs.
D. Input the parts cost, labor cost, tax cost, and the sublet repair cost into a calculator.

Answer A is incorrect. There are often more charges that need to be added in addition to labor and parts charges.

Answer B is correct. Using a computer that calculates all of the variables involved with a service repair is the most accurate method of figuring a total. It is also advisable to check the availability of parts prior to calling the customer.

Answer C is incorrect. The tax was not included in this list.

Answer D is incorrect. This list seems complete, but answer B is the best answer because it recommends using a computer program to calculate the total.

TASK B.3.3

36. Service Consultant A typically recommends that the alignment be checked prior to having new tires installed to assure longer lasting tire life. Service Consultant B only recommends the lowest price tires available in order to save the customer some expense. Who is correct?

A. A only
B. B only
C. Both A and B
D. Neither A nor B

Answer A is correct. Only Service Consultant A is correct. It is wise to recommend an alignment when new tires are being purchased. This helps promote longer tire life by having all of the wheels set at the correct angles.

Answer B is incorrect. The Service Consultant should have more than one choice of tires for the customer to choose from. She should be knowledgeable about the features and benefits of the various choices of tires available.

Answer C is incorrect. Only Service Consultant A is correct.

Answer D is incorrect. Service Consultant A is correct.

TASK A.1.12

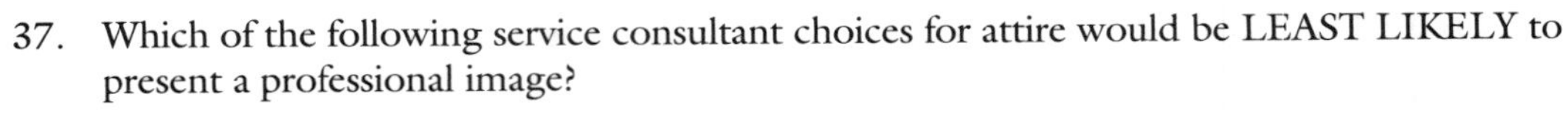

37. Which of the following service consultant choices for attire would be LEAST LIKELY to present a professional image?

A. Green shirt and blue pants
B. Red shirt and blue jeans
C. White shirt and tan pants
D. Clean and neatly groomed hair

Answer A is incorrect. This attire would be a good choice for a professional image.

Answer B is correct. Blue jeans are not considered a good choice for a professional image.

Answer C is incorrect. This attire would be a good choice for a professional image.

Answer D is incorrect. It is advisable to have clean and neatly groomed hair when attempting to present a professional image.

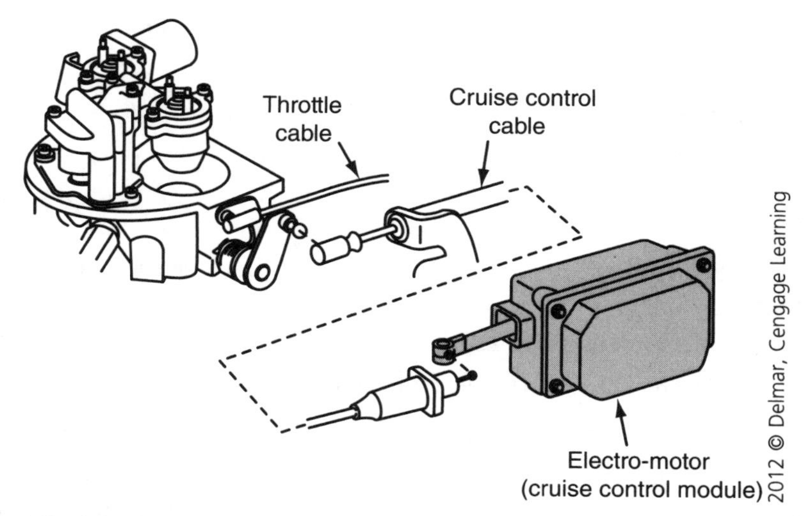

38. All of the following statements concerning the figure above are correct EXCEPT:

TASK B.4.3

A. The cruise control system must disengage when the brakes are applied.

B. The cruise control system must disengage when the clutch is applied.

C. The cruise control module is mounted to the cruise electromotor.

D. The cruise motor must have a vacuum hose connected in order to work properly.

Answer A is incorrect. All cruise control systems will disengage the system when the brake pedal is depressed. Some manufacturers include a set of contacts on the main brake switch for this purpose, while other manufacturers may have a separate brake switch to deactivate the cruise control system.

Answer B is incorrect. All cruise control systems will disengage the system when the clutch is depressed by including a clutch switch on the clutch pedal.

Answer C is incorrect. The cruise control system in the figure shows that the cruise control module mounts on the cruise motor.

Answer D is correct. The cruise motor in the figure operates with an electromotor and will not require a vacuum hose connection.

39. The refrigerant changes from a high-pressure vapor to a high-pressure liquid as it passes through which A/C component?

TASK B.4.2

A. Evaporator

B. Compressor

C. Receiver/drier

D. Condenser

Answer A is incorrect. The evaporator is the cold component of the A/C system that is located inside the duct assembly.

Answer B is incorrect. The compressor is the pump of the A/C system that compresses low pressure into high pressure.

Answer C is incorrect. The receiver/drier helps to dry out any moisture that may be inside the A/C system. It is located in the liquid line between the condenser and the thermal expansion valve.

Answer D is correct. The condenser is the heat exchanger that is located out in front of the radiator. The job of the condenser is to transfer some of the heat from the hot gaseous refrigerant into the outside air. This causes the refrigerant to change from a high-pressure gas to a high-pressure liquid.

TASK B.5.1

40. All of the following procedures would be performed during a typical 60,000-mile service EXCEPT:

    A. Replace the rear main seal.
    B. Replace the spark plugs.
    C. Replace the fuel filter.
    D. Inspect and adjust the brake system.

Answer A is correct. Replacing the rear main seal would not be included as part of a 60,000-mile service.

Answer B is incorrect. The spark plugs are usually replaced during a 60,000-mile service.

Answer C is incorrect. The fuel filter is typically replaced during a 60,000-mile service.

Answer D is incorrect. The brake system is typically inspected and adjusted during a 60,000-mile service.

TASK A.2.1

41. Service Consultant A takes very little time explaining the estimate due to the hurried pace of the repair shop. Service Consultant B calculates an estimate and then adds a small percentage to cover incidental costs that might arise. Who is correct?

    A. A only
    B. B only
    C. Both A and B
    D. Neither A nor B

Answer A is incorrect. Each customer deserves a thorough and accurate estimate. A hurried explanation will likely result in an angry customer.

Answer B is correct. Only Service Consultant B is correct. It is sometimes advisable to pad the repair estimate to cover incidental costs. If the repair actually comes in at a lower level, the customer will be pleasantly surprised.

Answer C is incorrect. Only Service Consultant B is correct.

Answer D is incorrect. Service Consultant B is correct.

TASK B.5.3

42. Service Consultant A says that it is wise to group several maintenance items into a package to increase the sense of good value to the customer. Service Consultant B says that a maintenance menu board should be posted in a convenient location near the service write-up area. Who is correct?

    A. A only
    B. B only
    C. Both A and B
    D. Neither A nor B

Answer A is incorrect. Service Consultant B is also correct.

Answer B is incorrect. Service Consultant A is also correct.

Answer C is correct. Both Service Consultants are correct. Grouping service items into packages increases the appeal to some customers who are looking to increase their spending power. It is also good business to post a service menu board near the service write-up area to prompt customers to inquire about services that are offered at that location.

Answer D is incorrect. Both Service Consultants are correct.

43. A vehicle is in the repair shop for an inspection. The technician recommends having the tires balanced for a vibration and the clock spring replaced due to an airbag light that is illuminated. The customer asks the service consultant to prioritize the two services in order of importance. Service Consultant A recommends the tire rotation since it is less expensive. Service Consultant B says that both items are of equal importance. Who is correct?

TASK A.2.2

A. A only
B. B only
C. Both A and B
D. Neither A nor B

Answer A is incorrect. The tire rotation would not be more important than the airbag repair.

Answer B is incorrect. The airbag repair would be a more important repair since it is a safety-related item.

Answer C is incorrect. Neither Service Consultant is correct.

Answer D is correct. Neither Service Consultant is correct. The airbag repair would be of higher importance since it is a safety-related item.

44. A vehicle is in the repair shop for a thermostat replacement. Service Consultant A recommends that the cooling system be totally flushed and refilled with new coolant while the thermostat is being replaced. Service Consultant B recommends that the cooling system be leak-tested with a pressure tester while the thermostat is being replaced. Who is correct?

TASK A.2.4

A. A only
B. B only
C. Both A and B
D. Neither A nor B

Answer A is incorrect. Service Consultant B is also correct.

Answer B is incorrect. Service Consultant A is also correct.

Answer C is correct. Both Service Consultants are correct. It is a good practice to recommend related repairs to customers in order to provide a high level of service and save them money. It would be wise to include coolant drain and fill, as well as a leak test of the cooling system, at the time that the thermostat is being replaced.

Answer D is incorrect. Both Service Consultants are correct.

45. All of the following services may be considered a sublet repair for some mechanical shops EXCEPT:

TASK C.2

A. Radiator repair
B. Air filter replacement
C. Painting a body panel
D. Transmission overhaul

Answer A is incorrect. Many mechanical shops are not equipped to repair radiators. This service is either sent out to a specialty shop or new radiators are installed.

Answer B is correct. A mechanical shop would never sublet an air filter replacement. Replacing the air filter is a very simple task that any mechanical shop could complete.

Answer C is incorrect. Mechanical shops are not usually equipped to paint body panels. This repair is always sent out to a body shop for completion.

Answer D is incorrect. Transmission overhaul repairs are often sent to a transmission specialty shop. Mechanical shops that do not overhaul transmissions will sometimes remove the transmission and then send it to a rebuilder for the overhaul. After the overhaul, the shop will then reinstall the unit.

TASK A.1.3

46. Service Consultant A says that you should directly quote the customer's comments on the repair order. Service Consultant B says that you should ask probing questions about when, where, and how the problem occurs and paraphrase the customer's comments on the repair order. Who is correct?

A. A only
B. B only
C. Both A and B
D. Neither A nor B

Answer A is incorrect. The service consultant needs to interpret what the customer is saying and record in writing what she hears in a coherent format. Most customers do not describe their problems in the terms that the technician would easily understand.

Answer B is correct. Only Service Consultant B is correct. A wise service consultant will ask good probing questions to find out when, where, and how the vehicle is malfunctioning. The consultant will then record meaningful descriptions of the problem onto the repair order.

Answer C is incorrect. Only Service Consultant B is correct.

Answer D is incorrect. Service Consultant B is correct.

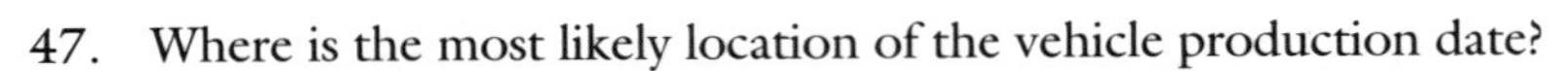

TASK B.7.2

47. Where is the most likely location of the vehicle production date?

A. On the B-pillar on the driver's side
B. In the trunk near the spare tire
C. On the B-pillar on the passenger's side
D. On the engine data tag

Answer A is correct. The vehicle production date is nearly always located on the B-pillar on the driver's side. A service consultant at a dealer has the option to retrieve the vehicle build date from the Dealer Support Network.

Answer B is incorrect. The vehicle production date is not usually in the trunk area. The vehicle options codes are sometimes located on a decal in the trunk.

Answer C is incorrect. The decal that contains this information is on the driver's side B-pillar.

Answer D is incorrect. The engine data tag will have detailed engine information but will not likely have the vehicle production date.

TASK A.2.1, A.2.6, A.3.4

48. Service Consultant A says that it is critical to have an accurate estimate before calling the customer for approval. Service Consultant B says that a reasonable completion time is important to have before calling for customer approval. Who is correct?

A. A only
B. B only
C. Both A and B
D. Neither A nor B

Answer A is incorrect. Service Consultant B is also correct.

Answer B is incorrect. Service Consultant A is also correct.

Answer C is correct. Both Service Consultants are correct. An accurate estimate and a reasonable completion time are both vital processes to have completed before calling the customer to receive approval to complete the repair.

Answer D is incorrect. Both Service Consultants are correct.

TASK A.1.6, A.1.12

49. All of the following would be normal activities in a well-run service facility EXCEPT:

    A. A customer enters the service write-up area and is quickly greeted by a service consultant.
    B. Technicians are working on customers' vehicles with several radios turned up loudly.
    C. Vehicle porters are putting seat covers, steering wheel covers, and floor mat covers into each car as it enters the service write-up area.
    D. Customers who are waiting for their vehicles have a clean and comfortable waiting area.

    Answer A is incorrect. Customers expect to be quickly greeted and processed when they arrive at a service facility.

    Answer B is correct. Having loud radios in a service facility does not provide a professional appearance to the customers who are there to receive service.

    Answer C is incorrect. Using the covers over the seats, steering wheels, and floor mats shows the customer that the facility cares about taking care of each vehicle.

    Answer D is incorrect. Providing a clean and comfortable waiting area is a good business practice for a service facility.

TASK C.3

50. Service Consultant A strives to keep an accurate customer appointment log by always writing the appointments down. Service Consultant B only schedules appointments to 50 percent of the shop's capacity so that the shop is not overbooked. Who is correct?

    A. A only
    B. B only
    C. Both A and B
    D. Neither A nor B

    Answer A is correct. Only Service Consultant A is correct. It is always advisable to write every appointment down in the appointment log in order to keep up with the expected customers each day.

    Answer B is incorrect. It is not a good idea to only book the shop to 50 percent of total capacity. This practice lowers shop profitability and does not encourage technician efficiency and productivity.

    Answer C is incorrect. Only Service Consultant A is correct.

    Answer D is incorrect. Service Consultant A is correct.

PREPARATION EXAM 3—ANSWER KEY

1. A
2. A
3. D
4. C
5. A
6. B
7. A
8. D
9. C
10. B
11. D
12. A
13. C
14. B
15. C
16. B
17. C
18. D
19. A
20. B
21. D
22. C
23. A
24. C
25. C
26. C
27. C
28. C
29. B
30. C
31. B
32. B
33. D
34. A
35. D
36. D
37. C
38. C
39. A
40. C
41. B
42. D
43. D
44. D
45. C
46. B
47. C
48. C
49. A
50. B

PREPARATION EXAM 3—EXPLANATIONS

TASK A.1.1

1. The telephone is ringing while a service consultant is working with a customer at the write-up desk. Service Consultant A does not answer the phone and lets the call go to voice mail. Service Consultant B answers the phone and engages in a 12-minute conversation. Who is correct?

A. A only
B. B only
C. Both A and B
D. Neither A nor B

Answer A is correct. Only Service Consultant A is correct. It would be acceptable to let the ringing phone go to voicemail if nobody was able to reach it. The key is to check the message as soon as possible and then return the call. If the Service Consultant does choose to answer the phone, then a message should be quickly taken to limit interruption to the customer.

Answer B is incorrect. It would not be advisable to answer the phone and have a long conversation in the middle of handling a customer in person.

Answer C is incorrect. Only Service Consultant A is correct.

Answer D is incorrect. Service Consultant A is correct.

2. All of these are components of the charging system EXCEPT:

TASK B.1.1

   A. Starter
   B. Serpentine belt
   C. Voltage regulator
   D. Alternator

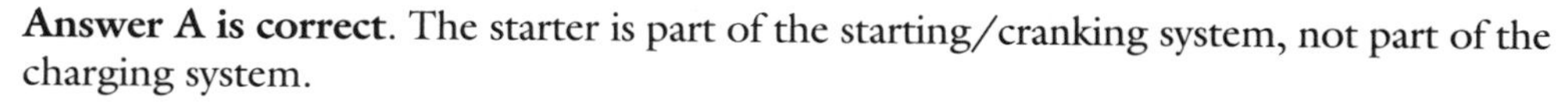

Answer A is correct. The starter is part of the starting/cranking system, not part of the charging system.

Answer B is incorrect. The belt drives the alternator.

Answer C is incorrect. The voltage regulator maintains system voltage and is a component of the charging system.

Answer D is incorrect. The alternator is the primary component of the charging system.

3. Service Consultant A says that providing a ballpark estimate is a useful tool for closing a sale. Service Consultant B says that asking for an appointment is a good way to close a sale. Who is correct?

TASK A.2.6

   A. A only
   B. B only
   C. Both A and B
   D. Neither A nor B

Answer A is incorrect. Ballpark estimates can potentially impede making sales due to their inaccurate nature. Accurate estimates should be researched and used when possible.

Answer B is incorrect. Asking for an appointment happens at a totally different time than at the point of closing a sale of a service to the customer.

Answer C is incorrect. Neither Service Consultant is correct.

Answer D is correct. Neither Service Consultant is correct. Closing the sale involves having a reasonable estimate that can be explained to the customer. The service consultant will be more effective if she has accurate data to relay to the customer.

4. Service Consultant A says that a sedan has four doors. Service Consultant B says that a coupe has two doors. Who is correct?

TASK B.7.4

   A. A only
   B. B only
   C. Both A and B
   D. Neither A nor B

Answer A is incorrect. Service Consultant B is also correct.

Answer B is incorrect. Service Consultant A is also correct.

Answer C is correct. Both Service Consultants are correct. Sedans have four doors and coupes have two doors. Other vehicle styles include convertibles, vans, sport utility vehicles, and crossover vehicles.

Answer D is incorrect. Both Service Consultants are correct.

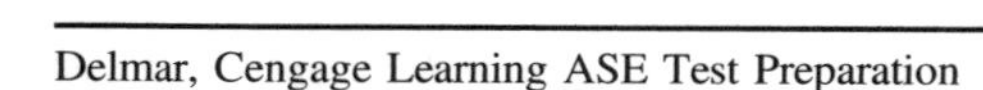

TASK A.1.5

5. Service Consultant A says that customers will sometimes ask for alternative transportation when they make an appointment. Service Consultant B says that most customers do not expect the service facility to provide some form of transportation if the service takes longer than they can readily wait. Who is correct?

A. A only
B. B only
C. Both A and B
D. Neither A nor B

Answer A is correct. Only Service Consultant A is correct. It is common for customers to ask for alternative transportation when the appointment is made. The service consultant needs to be aware that customers will seek this service and need to be prepared to handle these questions.

Answer B is incorrect. Many customers will not be satisfied to wait several hours while their vehicle is being repaired. Alternative transportation options need to be offered in those situations.

Answer C is incorrect. Only Service Consultant A is correct.

Answer D is incorrect. Service Consultant A is correct.

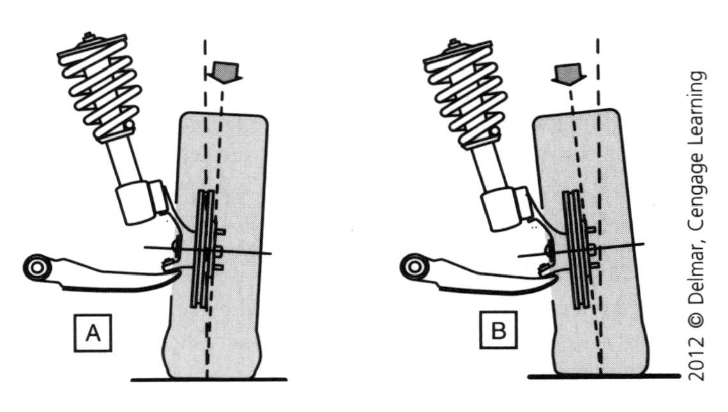

TASK B.3.3

6. Which wheel alignment angle is indicated in the figure?

A. Thrust angle
B. Camber
C. Caster
D. Toe

Answer A is incorrect. Thrust angle is the relationship between the overall toe direction of the rear wheels to the overall toe direction of the centerline of the chassis.

Answer B is correct. The amount the tire leans in or out at the top is called camber. When the tire leans out at the top, it is called positive camber. When it leans in at the top, it is called negative camber.

Answer C is incorrect. Caster is an imaginary line drawn through the ball joints or from the top of the strut mount to the lower ball joint.

Answer D is incorrect. Toe is the direction in which the front of the wheel points when going down the road.

7. Which of the following pieces of data can be determined from viewing the service history of a vehicle in the repair shop's computer system?

TASK A.1.7

A. A list of the service repair procedures performed at this location
B. Vehicle production date
C. The name of the salesperson who sold the car when it was new
D. A list of service repair procedures at other repair shops

Answer A is correct. The service history shows a list of the repair procedures performed at that service location. This information is helpful when planning current repair and maintenance for each vehicle.

Answer B is incorrect. The production date can be found on the B-pillar on the driver's side or it can be retrieved by a dealer from the dealer service network.

Answer C is incorrect. The selling location and salesperson for a vehicle would likely be available on a dealer service network inquiry.

Answer D is incorrect. There would not be a record of service performed at other repair shops in the service history of your shop.

8. A customer receives a letter from the manufacturer for which of these actions?

TASK B.6.3

A. A technical service bulletin (TSB)
B. A pattern failure
C. End of vehicle warranty
D. A manufacturer recall

Answer A is incorrect. TSBs are corrections to engineering mistakes; many may not be of strong importance.

Answer B is incorrect. Pattern failures are typically the reason that a TSB would get created. These types of documents are not sent out to the customers.

Answer C is incorrect. There is no need to announce by letter the end of a warranty period.

Answer D is correct. A manufacturer recall includes a mandatory letter to the vehicle's owner, since the recall usually involves repairs related to vehicle safety, use, or value.

9. Service Consultant A says that telling customers when their vehicle will be ready at the time they drop off the vehicle creates expectations. Service Consultant B says that accurate completion times can only be determined after vehicle inspection. Who is correct?

TASK A.1.9

A. A only
B. B only
C. Both A and B
D. Neither A nor B

Answer A is incorrect. Service Consultant B is also correct.

Answer B is incorrect. Service Consultant A is also correct.

Answer C is correct. Both Service Consultants are correct. Creating expectations before you know the real answers sets the whole service department up for failure, because you cannot know what is going to come up in diagnosis. This statement would exclude maintenance operations that you schedule by the hour and for which you can give accurate completion times. Service consultants should not make promises that are unrealistic. Get the answers first by diagnosing or inspecting the vehicle.

Answer D is incorrect. Both Service Consultants are correct.

TASK B.2.1

10. All of these systems use a filter EXCEPT:

    A. The automatic transmission
    B. The brake master cylinder
    C. The air conditioning (A/C) system
    D. The HVAC air handling system

Answer A is incorrect. Automatic transmissions most always utilize a filter. The filter is usually located inside the bottom transmission pan assembly.

Answer B is correct. Master cylinders do not have any kind of filter and are sealed from contaminants.

Answer C is incorrect. Some A/C systems use a filter to capture any small particles that could circulate and cause damage to the compressor or restrict refrigerant flow, thus causing cooling performance concerns.

Answer D is incorrect. Many late-model vehicles use a filter in the heating/air conditioning system to filter out particulates from the incoming air. Much like a furnace filter, cabin smoke and/or miscellaneous contaminants are filtered and trapped.

TASK A.1.13

11. What is the most likely benefit of performing a follow-up phone call after a repair visit?

    A. Seek referrals for new business from the customer.
    B. Offer discount coupons to the customer for them to give to friends and family.
    C. Offer discount repair services to the customer for returning to the shop for their next repair.
    D. Measure the satisfaction of the customer concerning the visit to the repair shop.

Answer A is incorrect. Seeking new referrals is the main benefit to performing follow-up calls.

Answer B is incorrect. Offering discounts to customers to share is not the main benefit to performing follow-up calls.

Answer C is incorrect. Offering discounts is not the main benefit to performing follow-up calls.

Answer D is correct. Follow-up calls are a valuable tool for a repair shop, because they allow the shop to measure the satisfaction of the customer's visit to the repair shop. They also show customers that the shop is interested in making sure that their needs and expectations were met.

TASK A.3.3, A.3.4

12. Service Consultant A always checks for parts availability before calling the customer with an estimate and completion time. Service Consultant B assumes that parts will be available and calls the customer in order to get the technician started on the repair as soon as possible. Who is correct?

    A. A only
    B. B only
    C. Both A and B
    D. Neither A nor B

Answer A is correct. Only Service Consultant A is correct. Checking on parts availability is a very important step in gathering an estimate and predicting a completion time.

Answer B is incorrect. It is not a good practice to take for granted that the necessary parts are going to be available for each vehicle.

Answer C is incorrect. Only Service Consultant A is correct.

Answer D is incorrect. Service Consultant A is correct.

13. Service Consultant A says that providing an estimate is required by law in some states. Service Consultant B says that explaining the details of the estimate helps to add value to the services the customer is buying from the shop. Who is correct?

TASK A.2.1

A. A only
B. B only
C. Both A and B
D. Neither A nor B

Answer A is incorrect. Service Consultant B is also correct.

Answer B is incorrect. Service Consultant A is also correct.

Answer C is correct. Both Service Consultants are correct. Some states have some sort of estimate law or motor vehicle repair act that requires shops to provide an estimate and receive approval before repairs are performed. Many times a sale is made on an item just because the consultant took the time to explain to the customer the value of having the repair done.

Answer D is incorrect. Both Service Consultants are correct.

14. The starter turns the engine by engaging with which of these components?

TASK B.1.2

A. Camshaft
B. Flywheel ring gear
C. Crankshaft
D. Battery

Answer A is incorrect. The camshaft provides a method of opening the intake and exhaust valves.

Answer B is correct. The starter drive gear engages the flywheel ring gear to rotate the engine for starting.

Answer C is incorrect. The crankshaft does turn during start up, but the starter turns it via the flywheel and its ring gear.

Answer D is incorrect. The battery is a storage device and does not engage or disengage the engine: It just provides the electrical power to do so.

15. Service Consultants need to be ready to answer questions from repair customers clearly. Service Consultant A provides clear and understandable answers to detailed questions that the customers have. Service Consultant B asks customers if they have any questions before completing the write-up process. Who is correct?

TASK A.2.3

A. A only
B. B only
C. Both A and B
D. Neither A nor B

Answer A is incorrect. Service Consultant B is also correct.

Answer B is incorrect. Service Consultant A is also correct.

Answer C is correct. Both Service Consultants are correct. Service consultants should be ready and willing to provide clear and understandable answers to customer questions. They should also ask the customer if they have any questions.

Answer D is incorrect. Both Service Consultants are correct.

TASK B.5.3

16. Service Consultant A says that maintenance schedules are not relevant, because it is impossible to estimate which driving style applies to the vehicle. Service Consultant B says that it is wise to follow the maintenance schedule that is closest to the driving style of each customer. Who is correct?

   A. A only
   B. B only
   C. Both A and B
   D. Neither A nor B

   Answer A is incorrect. Maintenance schedules are very useful when helping the customer make decisions about when to perform various services on their vehicle.

   Answer B is correct. Only Service Consultant B is correct. Maintenance schedules typically recommend different options, depending on how the vehicle is used. It is wise to ask the customer questions to try to match driving styles to the correct level of service.

   Answer C is incorrect. Only Service Consultant B is correct.

   Answer D is incorrect. Service Consultant B is correct.

TASK A.1.7, A.2.5

17. Repair shops work very hard to earn the trust of customers so that they will continue to patronize their location. Service Consultant A says that repeat customers receive more thorough service because all of the service records will be at one location. Service Consultant B says that repeat customers receive more thorough service because the repair technicians become familiar with the vehicle. Who is correct?

   A. A only
   B. B only
   C. Both A and B
   D. Neither A nor B

   Answer A is incorrect. Service Consultant B is also correct.

   Answer B is incorrect. Service Consultant A is also correct.

   Answer C is correct. Both Service Consultants are correct. Having all of the service records in one location, as well as the technicians' familiarity with the vehicle, are both good reasons to continue making all of the repairs at one location.

   Answer D is incorrect. Both Service Consultants are correct.

TASK A.1.3

18. A customer recites a list of symptoms to the service consultant. What is the most likely next step that the service consultant would do?

   A. Write down exactly what the customer says.
   B. Use his/her experience to estimate repairs.
   C. Offer suggestions about what the problem might be.
   D. Ask open-ended questions to determine customer needs.

   Answer A is incorrect. Customers are not technicians; ask questions and summarize customer responses.

   Answer B is incorrect. There is no need to give an estimate at this point; repair information should be given instead.

   Answer C is incorrect. Remain a consultant and don't try to be the technician. You must allow time for the technician to diagnose the problems.

   Answer D is correct. By asking the customers questions that help them flesh out their description of the problem, the service consultant can save both the customer and the technician time and money diagnosing the problem.

19. Providing accurate estimates is necessary for customers to make informed decisions about the repair of their vehicle. Service Consultant A carefully adds all of the charges together and then relays the information to the customer. Service Consultant B gives the customers a 25 percent discount in order to show the customer the value of the service being performed. Who is correct?

TASK A.2.1

A. A only
B. B only
C. Both A and B
D. Neither A nor B

Answer A is correct. Only Service Consultant A is correct. The service consultant should pay close attention to detail when putting together an estimate.

Answer B is incorrect. Discounts should not be used randomly in the normal activity of the shop. However, running service specials at various times during the year is a good way to increase car count in a repair shop.

Answer C is incorrect. Only Service Consultant A is correct.

Answer D is incorrect. Service Consultant A is correct.

20. Interpreting the technician's diagnosis is an important skill for a service consultant. Service Consultant A does not try to thoroughly understand each diagnosis a technician makes; he just provides vague explanations to the customers with whom he communicates. Service Consultant B is fairly knowledgeable about most vehicle systems and strives to interpret each technician diagnosis and then relay the information in a way that the customer will understand it. Who is correct?

TASK A.3.2

A. A only
B. B only
C. Both A and B
D. Neither A nor B

Answer A is incorrect. The more knowledgeable a service consultant is about the repair processes, the better he can communicate that to the customer. The service consultant should strive to continue educating himself as new systems evolve on new vehicles.

Answer B is correct. Only Service Consultant B is correct. It is important to give customers the most accurate and understandable explanation of their repairs as possible.

Answer C is incorrect. Only Service Consultant B is correct.

Answer D is incorrect. Service Consultant B is correct.

21. A customer brings a car to the repair shop with a complaint that the steering wheel vibrates while braking. Service Consultant A shows the customer all of the service coupon specials that are currently being offered by the shop. Service Consultant B tells the customer that the tires will need to be balanced to repair the problem. Who is correct?

TASK A.2.3

A. A only
B. B only
C. Both A and B
D. Neither A nor B

Answer A is incorrect. The first priority in this scenario would be to address the customer's original concern about a vibration while braking.

Answer B is incorrect. It is not advisable to attempt to diagnose the customer's vehicle during the write-up process.

Answer C is incorrect. Neither Service Consultant is correct.

Answer D is correct. Neither Service Consultant is correct. It is recommended that the service consultant stay on the task of addressing the customer's original concern before attempting to add additional services. It is also not advisable to attempt to diagnose the vehicle in the write-up area. The technicians are the experts in servicing the vehicles and they should be allowed to do their job.

TASK C.3

22. All of the following information should be on the customer appointment log EXCEPT:

    A. Customer name
    B. Estimated time of repair
    C. Vehicle color
    D. Vehicle year, make, and model

Answer A is incorrect. The customer name should always be on the appointment log.

Answer B is incorrect. A rough time of repair estimate should be on the customer appointment log. This time will usually not be extremely accurate, but it helps in planning the number of vehicles that need to be scheduled for each day.

Answer C is correct. The vehicle color is not a typical piece of information that would be on the customer appointment log.

Answer D is incorrect. The vehicle year, make, and model are usually present on the customer appointment log.

TASK B.3.1

23. The master cylinder is part of which system?

    A. Brake system
    B. Engine control system
    C. Automatic transmission system
    D. Power steering system

Answer A is correct. The master cylinder is the first component of the hydraulic portion of the brake system. When the brake pedal is depressed, it generates hydraulic pressure to apply the brake calipers and wheel cylinders.

Answer B is incorrect. The engine has many components, but none are referred to as the master cylinder.

Answer C is incorrect. The automatic transmission has many components, but none are referred to as the master cylinder.

Answer D is incorrect. The steering system has many components, but none are referred to as the master cylinder.

TASK A.2.2

24. Which of the following repairs would be the highest priority for a customer to repair in terms of the vehicle's safety?

    A. Replace the rear wiper blade.
    B. Service the automatic transmission fluid and filter.
    C. Replace a tire with a cut in the sidewall.
    D. Service an inoperative A/C system.

Answer A is incorrect. A worn rear wiper blade would not cause the vehicle to be unsafe to drive.

Answer B is incorrect. An automatic transmission that needs a service would not cause the vehicle to be unsafe to drive.

Answer C is correct. A tire with a cut in the sidewall is extremely dangerous, because it could blow out at any time. This is a very important repair that would need to be addressed immediately.

Answer D is incorrect. An inoperative A/C system would not cause the vehicle to be unsafe to drive.

25. A customer arrives to pick up their vehicle after extensive repairs are made. Service Consultant A calls the service porter to pull the vehicle around to the pickup area. Service Consultant B carefully explains all of the repairs and charges and asks the customer if there are any questions. Who is correct?

TASK A.1.14

   A. A only

   B. B only

   C. Both A and B

   D. Neither A nor B

   Answer A is incorrect. Service Consultant B is also correct.

   Answer B is incorrect. Service Consultant A is also correct.

   Answer C is correct. Both Service Consultants are correct. Much attention should be given to the customer as they pick up their vehicle. The porter should be called to bring the vehicle to the pickup area, and the service consultant should explain all of the repairs and charges to the customer.

   Answer D is incorrect. Both Service Consultants are correct.

26. Service Consultant A says that a technician's efficiency and speed should be monitored to determine how much work to schedule for him on the appointment log. Service Consultant B says that the quality of the technician's work should be monitored and communicated to management. Who is correct?

TASK A.3.5, A.3.7, A.3.8

   A. A only

   B. B only

   C. Both A and B

   D. Neither A nor B

   Answer A is incorrect. Service Consultant B is also correct.

   Answer B is incorrect. Service Consultant A is also correct.

   Answer C is correct. Both Service Consultants are correct. It is important to understand the efficiency of the technicians in the shop to assist in planning for work volume. It is also advisable to monitor the workmanship (quality) of the repairs that each technician produces.

   Answer D is incorrect. Both Service Consultants are correct.

27. All of the items below would be needed when collecting vehicle information for the vehicle repair order EXCEPT:

TASK A.1.2

   A. Vehicle identification number (VIN)

   B. Mileage

   C. Tire size

   D. Vehicle make, model, and color

   Answer A is incorrect. The VIN number is a very important piece of information to include on the vehicle repair order. The VIN has 17 digits that give information such as the manufacturer, the vehicle year, the engine, the body style, as well as the sequence in which the vehicle was built.

   Answer B is incorrect. The mileage of the vehicle should be put on the repair order. The repair order becomes a historical document for potential warranties and future repairs for the vehicle.

   Answer C is correct. The tire size may be needed for some repairs, but it is not a typical piece of information that would appear on every repair order.

   Answer D is incorrect. The vehicle make, model, and color should be present on the repair order to help the staff identify the vehicle while it is at the repair facility.

TASK B.1.1

28. Which item connects to the connecting rod at the end opposite the crankshaft?

    A. The flywheel
    B. The rear main seal
    C. The piston
    D. The cylinder head

Answer A is incorrect. The flywheel connects the crankshaft to the transmission.

Answer B is incorrect. The rear main seal prevents oil from leaking at the crankshaft rear.

Answer C is correct. A connecting rod connects to the piston on the top side and to the crankshaft on the bottom side.

Answer D is incorrect. The cylinder head is bolted to the cylinder block.

TASK A.2.2

29. A technician performs an extensive inspection and recommends the following: replacement of a damaged driver-side seat belt, cooling system flush, replacement of brake pads that have 7/32 inch remaining, and an oil change that is 1,800 miles overdue. Which of these represents the best way to prioritize this list to the customer?

    A. Brake pads, oil change, seat belt, cooling system service
    B. Seat belt, oil change, brake pads, cooling system service
    C. Seat belt, brake pads, oil change, cooling system service
    D. Oil change, cooling system service, brake pads, seat belt

Answer A is incorrect. The brake pads have no immediate need for replacement and are actually the third priority.

Answer B is correct. When a technician recommends a list of repairs, the service consultant must prioritize them based on need. The service consultant should always recommend addressing the safety items first, maintenance that is overdue next, and discretionary items that are not urgent last.

Answer C is incorrect. The oil change is 1,800 miles overdue, while the brakes still have life.

Answer D is incorrect. Safety items must be first and are the most immediate concern.

TASK B.1.3

30. Service Consultant A says that the thermostat is a component of the engine cooling system. Service Consultant B says that the heater core is a component of the engine cooling system. Who is correct?

    A. A only
    B. B only
    C. Both A and B
    D. Neither A nor B

Answer A is incorrect. Service Consultant B is also correct.

Answer B is incorrect. Service Consultant A is also correct.

Answer C is correct. Both Service Consultants are correct. The thermostat is a device that restricts coolant flow in the cooling system to speed the warm up process; it also maintains the correct engine temperature. The heater core is connected to the cooling system with hoses. The heater system blows air over the heater core to warm the duct air up to help warm the cab of the vehicle.

Answer D is incorrect. Both Service Consultants are correct.

31. Service Consultant A says that the torque converter is a component of the exhaust system. Service Consultant B says that the catalytic converter is a component of the exhaust system. Who is correct?

TASK B.1.3

A. A only
B. B only
C. Both A and B
D. Neither A nor B

Answer A is incorrect. The torque converter is a component of the automatic transmission. It serves as a fluid-coupling device that connects the engine to the transmission.

Answer B is correct. Only Service Consultant B is correct. The catalytic converter is a device used in the exhaust system that causes a chemical reaction to help reduce vehicle emissions.

Answer C is incorrect. Only Service Consultant B is correct.

Answer D is incorrect. Service Consultant B is correct.

32. All of these are components of an automatic transmission EXCEPT:

TASK B.2.1

A. Planetary gear sets
B. Pressure plate
C. Torque converter
D. Valve body

Answer A is incorrect. The planetary is a multiple-ratio gear set used in automatic transmissions to provide different gear ratios in a compact package.

Answer B is correct. The pressure plate is a component of the clutch in a manual transmission. All of the other items are components of an automatic transmission.

Answer C is incorrect. The torque converter is a hydraulic coupling device that connects the engine to the transmission. Most late-model automatic transmissions have an electronically controlled locking clutch inside the converter that makes a direct connection between the engine and the transmission during light-load cruising. This drops RPM and helps with gas mileage.

Answer D is incorrect. The valve body is the hydraulic control unit. Newer electronic transmissions have solenoids within the valve body that are run by the power train control module (PCM).

33. Service Consultant A does not groom her hair while at work. Service Consultant B wears clean casual clothes but does not tuck in his shirt while at work. Who is correct?

TASK A.1.12

A. A only
B. B only
C. Both A and B
D. Neither A nor B

Answer A is incorrect. A service consultant should be neatly groomed while at work in order to present a professional image.

Answer B is incorrect. A service consultant who does not tuck in his shirt is not practicing a professional image. The way that service consultant presents himself sends a message to the customer.

Answer C is incorrect. Neither Service Consultant is correct.

Answer D is correct. Neither Service Consultant is correct. Service consultants should groom themselves well, wear some type of professional clothing, and make sure it is clean, neat, and tucked in.

TASK B.2.2, B.2.3

34. An automatic transmission is being replaced. Service Consultant A says that the transmission cooler should be flushed prior to connecting to the transmission. Service Consultant B says that the engine oil must be changed to guarantee good transmission life. Who is correct?

A. A only
B. B only
C. Both A and B
D. Neither A nor B

Answer A is correct. Only Service Consultant A is correct. The transmission cooler must be flushed or replaced to remove debris from the old transmission. Debris from the failed transmission can cause problems with the replacement if not cleaned.

Answer B is incorrect. Transmissions do not share oil with the engine.

Answer C is incorrect. Only Service Consultant A is correct.

Answer D is incorrect. Service Consultant A is correct.

TASK A.3.1

35. A customer calls the repair shop to add a service request for a vehicle that is currently receiving service in the shop. Service Consultant A uses the intercom to relay the message to the technician. Service Consultant B sends a text message to the technician giving the customer's instructions. Who is correct?

A. A only
B. B only
C. Both A and B
D. Neither A nor B

Answer A is incorrect. Using the intercom system is not a recommended way to communicate with a technician about an added service to a repair order.

Answer B is incorrect. Sending a text message is not a recommended way to communicate with a technician about an added service to a repair order.

Answer C is incorrect. Neither Service Consultant is correct.

Answer D is correct. Neither Service Consultant is correct. The service consultant should personally add the additional service to the repair order and then advise the technician of the additional service.

TASK B.3.2

36. Which of the following type of brake fluid has the lowest boiling point?

A. DOT 4
B. DOT 5
C. DOT 5.1
D. DOT 3

Answer A is incorrect. DOT 4 brake fluid is used on some vehicles and has a boiling point of about 446°F (230°C).

Answer B is incorrect. DOT 5 brake fluid is used on some high-performance vehicles and has a boiling point of 500°F (260°C). This fluid should never be recommended for use on a vehicle that has ABS brakes.

Answer C is incorrect. DOT 5.1 brake fluid is used on some high-performance vehicles and has a boiling point of 500°F (260°C).

Answer D is correct. DOT 3 is used on a high percentage of vehicles and has the lowest boiling point of 401°F (205°C).

37. Which of the following is most likely to have the greatest impact on a customer's decision to do business with a repair shop?

TASK A.1.6

A. Extended business hours
B. The service consultant's appearance
C. The level of trust they feel
D. Discount pricing

Answer A is incorrect. Extended hours may be a nice convenience for a customer, but only a very small number would do business with a repair shop solely because of this practice.

Answer B is incorrect. Appearance is very important, but it is not the biggest factor that influences customers to visit a repair shop.

Answer C is correct. When a customer trusts a repair shop, he or she is more likely to do business there.

Answer D is incorrect. It is possible that some customers will do business with a shop based on discount prices, but they tend to have no loyalty to the shop and will go elsewhere if the deal is better. Statistically, price is less a factor than is trust and how the customer is treated. If there are issues with the facility or the repair, the lower price is quickly forgotten.

38. Which of these is a component of the supplemental restraint system (SRS) or airbag system?

TASK B.4.1

A. Throttle sensor
B. Power train control module (PCM)
C. Clock spring/spiral cable
D. Wheel speed sensor

Answer A is incorrect. A throttle sensor is an engine management input device.

Answer B is incorrect. A PCM is a component of an engine management system.

Answer C is correct. The clock spring or spiral cable keeps constant electrical contact between the inflator module in the steering wheel and the airbag module when the steering wheel turns.

Answer D is incorrect. A wheel speed sensor is part of the antilock brake system.

39. All of the following components are needed to operate the electric horn EXCEPT:

TASK B.4.3

A. Horn computer
B. Horn relay
C. Horn switch
D. Clock spring

Answer A is correct. It is not typical to have a computer that just operates the horn functions.

Answer B is incorrect. Most vehicles use a horn relay to energize the horn when the switch is depressed.

Answer C is incorrect. All horns have some type of switch to depress in order to activate the relay.

Answer D is incorrect. Vehicles that have airbags use a clock spring to connect electrical power to the rotating steering wheel. The horn circuit is also included in the clock spring.

TASK B.4.2

40. The vehicle climate control system creates passenger compartment heat by which method?

A. An electric heater strip inside the dash
B. A microwave heating grid inside the dash
C. A heat exchanger, called a heater core, inside the dash
D. A heat exchanger, called the evaporator, inside the dash.

Answer A is incorrect. Automotive heaters typically do not use electric heat strips to create heat.

Answer B is incorrect. Automotive heaters typically do not use microwave heating grids to create heat.

Answer C is correct. Hot water is pumped from the engine into a small heat exchanger called a heater core that is located in the HVAC duct assembly. The blower motor moves air past the heater core when the controls are set for heat. The air is then directed at the defrost, vent, or heater ducts.

Answer D is incorrect. The evaporator is the heat exchanger for the A/C system. It is the device that makes the air cold when the compressor is engaged.

TASK B.5.1

41. Which of the following procedures would likely be performed during a typical 60,000-mile service?

A. Replace the axle bearings.
B. Drain and fill the transmission fluid.
C. Replace the intake manifold gaskets.
D. Drain and fill the windshield washer solvent.

Answer A is incorrect. Axle bearing replacement is not part of a typical 60,000-mile service.

Answer B is correct. Draining and filling the transmission fluid is part of a typical 60,000-mile service. Other items included in a 60,000-mile service typically are replacing the engine filters, draining and filling the engine and drive train fluids, inspecting and adjusting the brake system, and performing an overall inspection of most of the systems on a vehicle.

Answer C is incorrect. Replacing the intake manifold gasket is not part of a typical 60,000-mile service.

Answer D is incorrect. Draining the windshield solvent is not part of a 60,000-mile service.

TASK B.5.2

42. Each of the following procedures would be performed during a typical oil change service EXCEPT:

A. Drain and refill the engine oil.
B. Replace the oil filter.
C. Check the basic fluid levels and tire pressures.
D. Replace the air filter.

Answer A is incorrect. The oil is always drained and refilled during an oil change.

Answer B is incorrect. The oil filter is typically replaced during an oil change.

Answer C is incorrect. The basic fluid levels and tire pressures are checked during an oil change.

Answer D is correct. The air filter is typically checked during an oil change, but it is not always replaced. If the technician recommends replacing it, then approval from the customer needs to be attained.

43. All of the following are benefits of recommending additional services to a current customer who has left his/her car at your shop EXCEPT:

TASK A.2.4

A. The customer will begin to trust that the shop is looking out for his/her well-being and safety.
B. The shop is more profitable from the increased sales of parts.
C. The shop is more profitable from the increased sales of labor.
D. The customer will compare the prices of the services with other shops.

Answer A is incorrect. Customers appreciate it when service professionals make honest recommendations on needed service for their vehicles.

Answer B is incorrect. Additional services that a shop performs have the potential to increase the shop's profits by increasing parts sales.

Answer C is incorrect. A shop that performs more services will obviously make more income due to the additional labor charges.

Answer D is correct. Recommending additional services to customers will not cause them to compare your prices with other shops.

44. Which items would be typically covered under a manufacturer's power train warranty?

TASK B.6.2, B.6.3, B.6.4

A. Belts and hoses
B. Wheels and tires
C. Front-end and suspension items
D. Engine and transmission

Answer A is incorrect. Belts and hoses are considered wear items that are rarely covered under a power train warranty. These items would typically only be covered for manufacturer's defects for a short time after the vehicle is purchased new.

Answer B is incorrect. Wheels and tires are considered wear items and would not be covered under a power train warranty.

Answer C is incorrect. Front-end and suspension items might be covered under a bumper-to-bumper warranty for the duration of a specified combination of time and mileage. Frequently, these warranties are for three to five years and from 36,000 miles to 50,000 miles. Actual warranties vary among the various manufacturers.

Answer D is correct. The engine and transmission are typically the systems covered under a power train warranty.

45. Which of the following services would be included as part of a 30,000-mile service?

TASK A.1.8, B.5.3

A. Exhaust gasket replacement
B. Intake manifold gasket replacement
C. Fuel filter replacement
D. Hub bearings repacking

Answer A is incorrect. A 30,000-mile service does not typically include an exhaust gasket replacement.

Answer B is incorrect. A 30,000-mile service does not typically include an intake manifold gasket replacement.

Answer C is correct. A 30,000-mile service typically includes replacement of all of the engine filters and fluids as well as performing tests and inspections and tests on many other maintenance items.

Answer D is incorrect. Hub bearings are sealed and do not ever need to be repacked.

TASK B.7.2

46. Which of the following is the most likely location to find the vehicle production date?

    A. Stamped on the valve cover
    B. On the driver's side B-pillar
    C. The emission decal under the hood
    D. Inside the gas door

Answer A is incorrect. Dates stamped on valve covers could be the day the engine was built or even the day the valve cover was made.

Answer B is correct. There are other possible locations, but this is the only correct option in the list of choices provided.

Answer C is incorrect. The emissions sticker is often on the radiator support or under the hood.

Answer D is incorrect. Some manufacturers put their calibration code on the gas door, but usually the gas door just contains instructions about opening the cap and fueling the vehicle.

TASK A.1.4

47. Service Consultant A says that when greeting a customer, you should offer your name and a handshake. Service Consultant B says that when greeting a customer, the service consultant should make eye contact and smile when welcoming them. Who is correct?

    A. A only
    B. B only
    C. Both A and B
    D. Neither A nor B

Answer A is incorrect. Service Consultant B is also correct.

Answer B is incorrect. Service Consultant A is also correct.

Answer C is correct. Both Service Consultants are correct. Cordial greetings should be exchanged when welcoming a customer. Shaking the customer's hand and making eye contact are a way to make the customer feel welcome.

Answer D is incorrect. Both Service Consultants are correct.

TASK C.2

48. Service Consultant A says that some repair shops have to sublet body repairs to a body repair shop. Service Consultant B says that the charges for a sublet repair should be included on the repair order. Who is correct?

    A. A only
    B. B only
    C. Both A and B
    D. Neither A nor B

Answer A is incorrect. Service Consultant B is also correct.

Answer B is incorrect. Service Consultant A is also correct.

Answer C is correct. Both Service Consultants are correct. It is common to sublet body repairs out to a body shop. It is important to include the charges for any sublet repairs on the repair order so the service facility can collect the right amount for each vehicle.

Answer D is incorrect. Both Service Consultants are correct.

49. A customer arrives at the repair shop requesting an oil change and a tire rotation. Service Consultant A recommends that the shop perform a brake cleaning and inspection while the wheels are removed. Service Consultant B recommends that the fuel system be serviced while the vehicle is at the shop. Who is correct?

TASK A.2.4

A. A only
B. B only
C. Both A and B
D. Neither A nor B

Answer A is correct. Only Service Consultant A is correct. It is a good practice to educate the customer about related services on their vehicle. This shows the customer that the service consultant is trying to provide thorough service as well as save them money and time by grouping services when possible.

Answer B is incorrect. A fuel-system service would not be considered a service that is related to an oil change and tire rotation.

Answer C is incorrect. Only Service Consultant A is correct.

Answer D is incorrect. Service Consultant A is correct.

50. A vehicle that is in the shop for a cooling system repair needs to have a radiator tank replaced. Service Consultant A recommends that the radiator be replaced with a new unit since the shop is not equipped to make the radiator repair. Service Consultant B recommends sending the radiator to a radiator specialty shop to have the repair completed. Who is correct?

TASK C.2

A. A only
B. B only
C. Both A and B
D. Neither A nor B

Answer A is incorrect. It would likely be more appropriate to have a tank installed at a radiator shop than to replace the whole radiator.

Answer B is correct. Only Service Consultant B is correct. It is common to sublet out some repairs to specialty shops if the general repair shop does not have the capacity to do it.

Answer C is incorrect. Only Service Consultant B is correct.

Answer D is incorrect. Service Consultant B is correct.

PREPARATION EXAM 4—ANSWER KEY

1. C
2. A
3. A
4. A
5. C
6. A
7. B
8. D
9. B
10. B
11. B
12. B
13. C
14. A
15. D
16. D
17. D
18. B
19. C
20. C
21. A
22. C
23. D
24. A
25. A
26. D
27. C
28. D
29. A
30. C
31. C
32. A
33. B
34. A
35. B
36. A
37. B
38. B
39. A
40. B
41. C
42. C
43. B
44. C
45. C
46. D
47. D
48. D
49. A
50. B

PREPARATION EXAM 4—EXPLANATIONS

TASK A.1.12, A.3.5

1. Service Consultant A encourages his/her technicians to keep their work areas as clean and orderly as possible. Service Consultant B continually monitors the status of the repair work that is being performed and updates the customers if problems arise. Who is correct?

A. A only

B. B only

C. Both A and B

D. Neither A nor B

Answer A is incorrect. Service Consultant B is also correct.

Answer B is incorrect. Service Consultant A is also correct.

Answer C is correct. Both Service Consultants are correct. Service consultants should strive to encourage the technicians to keep their work area neat. In addition, the service consultant should be in constant communication with the technician to monitor the status of the repair work.

Answer D is incorrect. Both Service Consultants are correct.

2. All of the following pieces of information are needed during the service write-up process EXCEPT:

TASK A.1.2

   A. Driver's license number for the customer
   B. Vehicle year, make, and model
   C. Customer's name
   D. List of services to be addressed

Answer A is correct. The customer driver's license number is not a typical piece of information that is needed during the service write-up process.

Answer B is incorrect. The vehicle year, make, and model are pieces of information that are always included on the repair order.

Answer C is incorrect. The customer's name always appears on the repair order.

Answer D is incorrect. A list of services requested is typically something that is defined in the conversation between the customer and service consultant during the write-up process.

3. Service Consultant A says a technical service bulletin (TSB) is issued when pattern failures happen to a certain type of vehicle. Service Consultant B says that vehicle manufacturers always cover the expenses associated with vehicles affected by TSBs. Who is correct?

TASK B.6.2

   A. A only
   B. B only
   C. Both A and B
   D. Neither A nor B

Answer A is correct. Only Service Consultant A is correct. Most manufacturers issue TSBs to assist technicians in repairing pattern failures. The manufacturer would likely cover such conditions if the vehicle is still under warranty coverage.

Answer B is incorrect. If the vehicle is out of warranty, then the customer will usually have to pay to for procedures performed that are the topic of technical service bulletins.

Answer C is incorrect. Only Service Consultant A is correct.

Answer D is incorrect. Service Consultant A is correct.

4. Which of the following practices would be the LEAST LIKELY method for suitable alternative transportation?

TASK A.1.5

   A. Explaining the location of the bus stop
   B. Taking the customer home
   C. Setting up a rental car
   D. Arranging a free loaner car

Answer A is correct. Sending a repair customer to the bus stop would not typically be a suitable form of alternate transportation.

Answer B is incorrect. Taking the customer home is a possible option as suitable transportation.

Answer C is incorrect. Setting up a rental car is a possible option as suitable transportation. Some extended warranty companies provide rental car coverage for covered repairs.

Answer D is incorrect. Some auto repair facilities provide loaner cars for repair customers.

TASK B.1.3

5. Service Consultant A says that the water pump is a component of the engine cooling system. Service Consultant B says that the catalytic converter is a component of the emissions system. Who is correct?

   A. A only
   B. B only
   C. Both A and B
   D. Neither A nor B

   Answer A is incorrect. Service Consultant B is also correct.

   Answer B is incorrect. Service Consultant A is also correct.

   Answer C is correct. Both Service Consultants are correct. The water pump is a component of the engine cooling system and the catalytic converter is a component of the emissions system.

   Answer D is incorrect. Both Service Consultants are correct.

TASK A.1.7

6. Service Consultant A says that viewing the service history on a vehicle can show which technicians have performed repairs on a vehicle. Service Consultant B says that viewing the service history on a vehicle can show when the vehicle was delivered to the selling dealership. Who is correct?

   A. A only
   B. B only
   C. Both A and B
   D. Neither A nor B

   Answer A is correct. Only Service Consultant A is correct. The service history is typically stored in an electronic format and gives details such as the repair procedures performed, the dates, and the technician who performed each repair.

   Answer B is incorrect. The service history would not reveal the dates of arrival at the selling dealership.

   Answer C is incorrect. Only Service Consultant A is correct.

   Answer D is incorrect. Service Consultant A is correct.

TASK C.1

7. A customer walks into the repair shop at 4:35 p.m. requesting an oil change and a cabin air filter replacement. The estimated time for this repair is 45 minutes and the shop closes at 5:00 p.m. Service Consultant A quickly completes the repair order and promises that the car will be completed that day before the close of business. Service Consultant B recommends that the customer reschedule this repair due to the limited time left in the day. Who is correct?

   A. A only
   B. B only
   C. Both A and B
   D. Neither A nor B

   Answer A is incorrect. This 45-minute repair will not be completed by the close of business. This customer should be scheduled for an appointment as soon as the schedule allows for it.

   Answer B is correct. Only Service Consultant B is correct. This job will run longer than the shop stays open, so the likelihood of something going wrong is increased. The best option is to schedule this operation at the next convenient time that the schedule allows.

   Answer C is incorrect. Only Service Consultant B is correct.

   Answer D is incorrect. Service Consultant B is correct.

8. What is the most likely method to find an accurate repair procedure status?

TASK A.1.9, A.3.5

A. Ask the service manager to view the activity and progress.
B. Walk through the shop to view the activity and progress.
C. Ask the car porter to go inspect the technician's progress.
D. Physically go and discuss the progress with the technician.

Answer A is incorrect. The service manager is not likely to know the status of the repair procedures that are going on in the shop at any given time.

Answer B is incorrect. An accurate status of repair procedures is not possible to attain by just walking through the shop.

Answer C is incorrect. The car porter is not a good option to use when trying to find out the status of a repair procedure. The service consultant needs to go and investigate the situation.

Answer D is correct. The best way to determine the status of a repair job is to go and discuss the progress with the technician.

9. Which type of oil change reminder used on some late-model vehicles must be reset after each oil change?

TASK B.5.2

A. Oil pressure gauge
B. Change oil indicator
C. Oil pressure light
D. Oil life chime

Answer A is incorrect. The oil pressure gauge does not need to be reset after an oil change. This device provides live engine oil pressure information to the driver when the engine is running.

Answer B is correct. The change oil indicator light or message in the information center needs to be reset after each oil change.

Answer C is incorrect. The oil pressure light comes on with the key in the accessory position. It also comes on if the engine is running and loses oil pressure to warn the driver to shut the engine down.

Answer D is incorrect. Most vehicles do not have any type of audible reminder for oil change. The reminder is typically a light or a message in the information center.

10. Service consultants need to present a professional image while at work. Service Consultant A wears flip-flop shoes with no socks to work. Service Consultant B wears a light blue uniform shirt with tan shorts to work. Who is correct?

TASK A.1.12

A. A only
B. B only
C. Both A and B
D. Neither A nor B

Answer A is incorrect. Flip-flop shoes are not a good choice for a professional image as a service consultant.

Answer B is correct. Only Service Consultant B is correct. Professional choice of dress is very important for service consultants to make a good impression. The service consultant is the person that the customers encounter most frequently, so a good image is important.

Answer C is incorrect. Only Service Consultant B is correct.

Answer D is incorrect. Service Consultant B is correct.

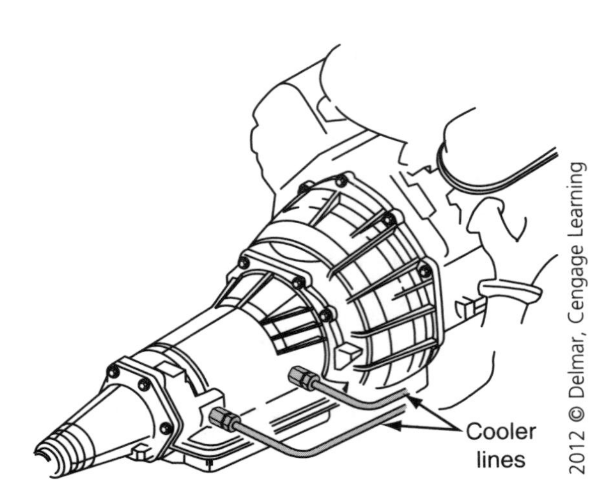

TASK B.2.2

11. What purpose do the cooler lines in the figure above serve on an automatic transmission?

    A. To connect the transmission to the engine oil pan-mounted transmission cooler
    B. To connect the transmission to the radiator-mounted transmission cooler
    C. To connect the transmission coolant to the radiator
    D. To connect the transmission to the air-to-air transmission heater

    Answer A is incorrect. The transmission cooler is not mounted in the engine oil pan.

    Answer B is correct. The cooler lines connect the transmission to the radiator-mounted transmission cooler assembly. Some vehicles have an additional transmission oil cooler that is mounted in front of the radiator.

    Answer C is incorrect. The transmission does not use coolant; it uses automatic transmission fluid.

    Answer D is incorrect. The automatic transmission does not use a device called an air-to-air heater.

TASK A.1.14, B.4.2, B.6.1

12. A customer arrives to pick up his/her vehicle after the A/C compressor was replaced. After paying the bill, the customer comes back to the service desk to ask why the accumulator was replaced. Service Consultant A explains that all vehicles that have any type of A/C repairs need to have an accumulator replaced. Service Consultant B says that the accumulator should always be replaced whenever the compressor is replaced. Who is correct?

    A. A only
    B. B only
    C. Both A and B
    D. Neither A nor B

    Answer A is incorrect. The accumulator does not have to be replaced after some A/C repairs that do not cause the system to be exposed to contaminants.

    Answer B is correct. Only Service Consultant B is correct. Replacing the accumulator is a vital step when the A/C compressor is replaced. Many compressor manufacturers will not honor the compressor warranty unless the accumulator is replaced at the time of repair.

    Answer C is incorrect. Only Service Consultant B is correct.

    Answer D is incorrect. Service Consultant B is correct.

13. A vehicle is in the repair shop with a complaint of poor heater performance. Service Consultant A says that the engine cooling system may need to be diagnosed. Service Consultant B says that a stuck heater control valve could be the cause. Who is correct?

TASK B.4.3

A. A only
B. B only
C. Both A and B
D. Neither A nor B

Answer A is incorrect. Service Consultant B is also correct.

Answer B is incorrect. Service Consultant A is also correct.

Answer C is correct. Both Service Consultants are correct. The heater system on a vehicle works in conjunction with the engine cooling system, so it would need to be checked when there is a heater concern. Many heater systems use a heater control valve, which regulates hot coolant into the heater core when the temperature control lever is moved to the heat position.

Answer D is incorrect. Both Service Consultants are correct.

14. A vehicle with 115,250 miles on the odometer is in for service at a repair center for a problem of the heater not getting warm. The diagnosis is a restricted heater core. Service Consultant A prepares an estimate to replace the heater core. Service Consultant B prepares an estimate to replace the heater core, heater hoses, thermostat, and the radiator cap. Who is correct?

TASK A.2.2

A. A only
B. B only
C. Both A and B
D. Neither A nor B

Answer A is correct. Only Service Consultant A is correct. It is correct to give an estimate of correcting the problem of a vehicle. It is also advisable to have an estimate of additional repairs that could be done at the same time in order to thoroughly service the vehicle.

Answer B is incorrect. It is not a good idea to recommend significant additional services in the primary estimate. However, it is a good practice to be ready to up-sell the repair with additional related services.

Answer C is incorrect. Only Service Consultant A is correct.

Answer D is incorrect. Service Consultant A is correct.

15. A vehicle is in the repair shop for a problem of the engine stalling when coming to a stop. The technician diagnoses the problem as a dirty throttle body and recommends a throttle-body cleaning service. Service Consultant A says that this repair will take several hours to complete. Service Consultant B says that this service operation involves removing the cylinder heads. Who is correct?

TASK A.3.2

A. A only
B. B only
C. Both A and B
D. Neither A nor B

Answer A is incorrect. Servicing the throttle body does not take several hours to complete. It can usually be completed in about one hour on many vehicle lines.

Answer B is incorrect. The cylinder heads would not have to be removed to clean the throttle body.

Answer C is incorrect. Neither Service Consultant is correct.

Answer D is correct. Neither Service Consultant is correct. Cleaning the throttle body is a common repair that is performed on fuel injection engines. This service is considered a maintenance-related service that can be completed on many vehicles in about one hour.

TASK A.2.4

16. Which of the following car services is an operation that could be performed at the same time as an oil pan gasket replacement?

A. Cabin air filter replacement
B. Transmission fluid and filter service
C. Intake manifold gasket
D. Oil and filter change

Answer A is incorrect. The cabin air filter is not a service that is related to the oil pan gasket.

Answer B is incorrect. Performing a transmission fluid and filter service would not align with replacing the oil pan gasket.

Answer C is incorrect. Repairing the intake manifold gasket is not closely related to repairing the oil pan gasket.

Answer D is correct. Changing the oil and filter is a service that could be performed at the same time as the oil pan gasket.

TASK B.4.1

17. Service Consultant A says that the airbag clock spring is located at the bottom of the steering column. Service Consultant B says that the airbag inflator module should be stored face down in a safe location while service is being performed. Who is correct?

A. A only
B. B only
C. Both A and B
D. Neither A nor B

Answer A is incorrect. The airbag clock spring is located at the top of the steering column. This component allows the steering wheel to rotate while providing a hard-wired connection to the airbag that is mounted on the steering wheel.

Answer B is incorrect. The airbag inflator module should not be stored face down due to the possibility of an inadvertent deployment, which would cause it to be projected upward rapidly.

Answer C is incorrect. Neither Service Consultant is correct.

Answer D is correct. Neither Service Consultant is correct. The clock spring is located at the top of the steering wheel. Airbag inflator modules should be stored face up for safety purposes.

TASK A.2.6, A.3.3

18. Service Consultant A says that an accurate estimate is not important to have before calling the customer for approval. Service Consultant B says that the availability of the parts should be checked prior to calling the customer for approval. Who is correct?

A. A only
B. B only
C. Both A and B
D. Neither A nor B

Answer A is incorrect. It is not a good idea to call the customer prior to assembling a fairly accurate estimate that includes finding out the availability of the repair parts.

Answer B is correct. Only Service Consultant B is correct. It is very important to find out the availability of the repair parts prior to calling the customer.

Answer C is incorrect. Only Service Consultant B is correct.

Answer D is incorrect. Service Consultant B is correct.

19. Service Consultant A always provides clear estimates to her customers so they can make informed decisions about their vehicle. Service Consultant B always checks on the parts warranty in order to inform the customer about this detail. Who is correct?

TASK A.2.1, B.6.2

A. A only
B. B only
C. Both A and B
D. Neither A nor B

Answer A is incorrect. Service Consultant B is also correct.

Answer B is incorrect. Service Consultant A is also correct.

Answer C is correct. Both Service Consultants are correct. Providing clear estimates after checking on the availability of parts, as well as any warranties involved, is a good practice for a service consultant to follow. This gives him or her the best opportunity to sell the repair service to the customer.

Answer D is incorrect. Both Service Consultants are correct.

20. Which of the following repair procedures would be the most important repair to be performed as related to vehicle safety?

TASK A.2.2

A. Wheel alignment
B. Rear window defogger
C. Brake line replacement
D. Power window motor

Answer A is incorrect. A wheel alignment would not be a safety-related repair procedure. Having properly aligned wheels will improve the handling and tire-wear characteristics of the vehicle.

Answer B is incorrect. A rear window defogger repair would not be a critical safety-related repair.

Answer C is correct. A brake line replacement is a very critical vehicle safety concern. Any safety-related repair should always be at the forefront of recommended repairs to a customer.

Answer D is incorrect. A power window motor repair would not be a safety-related repair.

21. Service Consultant A says that a three-year free replacement is a warranty feature on some premium batteries. Service Consultant B says that a lifetime guarantee is a warranty feature on some premium batteries. Who is correct?

TASK A.2.5, B.6.1

A. A only
B. B only
C. Both A and B
D. Neither A nor B

Answer A is correct. Only Service Consultant A is correct. There are a few battery manufacturers who offer a three-year free replacement feature on their batteries. Being aware of the parts warranties is important for good service consultants.

Answer B is incorrect. There are no battery manufacturers who offer a lifetime warranty on their products. Batteries have a limited lifespan.

Answer C is incorrect. Only Service Consultant A is correct.

Answer D is incorrect. Service Consultant A is correct.

TASK A.2.6

22. Which of the following reasons could a service consultant use to finalize a service repair sale?

A. The fuel economy will be reduced if the maintenance is not performed on time.

B. The owner will be more likely to trade the vehicle because she is dissatisfied with it.

C. The vehicle will be more reliable if the correct service and maintenance are performed on time.

D. It is cheaper to drive a vehicle if maintenance is ignored.

Answer A is incorrect. A vehicle that is properly maintained will exhibit improved performance and fuel economy.

Answer B is incorrect. Having to trade the vehicle is not a positive reason to give for performing needed services.

Answer C is correct. A well-maintained vehicle will provide more reliable service than one that is poorly maintained.

Answer D is incorrect. A well-maintained vehicle will cost less to own than one that is poorly maintained due to the increased likelihood of untimely breakdowns.

TASK A.3.1

23. All of the following methods of communicating a customer request to a technician are common EXCEPT:

A. Paging the technician to the write-up desk to relay the message in person.

B. Writing the message down and having the car porter give the message to the technician.

C. Walking back to the technician and adding the customer request to the repair order.

D. Having a car porter call the technician to relay the message.

Answer A is incorrect. This one-on-one method would be an acceptable way to relay customer requests to the technician.

Answer B is incorrect. Putting the message in a written form and then sending the message to the technician would be an acceptable way to get a message to the technician.

Answer C is incorrect. Walking back and directly putting the customer request on the repair order is an acceptable way.

Answer D is correct. Having a porter call the technician would likely cause parts of the message to be lost in the chain of people communicating its contents.

TASK A.2.3, A.3.2

24. A customer comments that his car pulls to the left while driving on a level road. Service Consultant A schedules the repair with the suspension specialist. Service Consultant B thinks that the problem could be unequal tire pressure and offers to check the tire pressure for free. Who is correct?

A. A only

B. B only

C. Both A and B

D. Neither A nor B

Answer A is correct. Only Service Consultant A is correct. The service consultant's job is to get the vehicle to the technician so a diagnosis can be made.

Answer B is incorrect. It is not a good idea for the service consultant to attempt to diagnose the vehicle in the write-up lane. He should leave the diagnosis to the technicians.

Answer C is incorrect. Only Service Consultant A is correct.

Answer D is incorrect. Service Consultant A is correct.

25. Which of the following activities would be the most likely method of promoting open communication among the repair shop employees?

TASK A.3.8

    A. Weekly meetings that allow all employees to give input to the operation
    B. Putting a suggestion box in the employee break area
    C. Having a yearly Christmas party at a local restaurant
    D. Sending birthday cards to employees on their birthday

    Answer A is correct. A shop that has regular opportunities for employees to meet and receive training, as well as voice concerns, is a positive way to promote open communication.

    Answer B is incorrect. A suggestion box in the employee break area would be one way to promote communication, but direct communication would be better.

    Answer C is incorrect. Having a yearly Christmas party is a positive practice for employees to engage in, but it would not promote constant communication throughout the year.

    Answer D is incorrect. Sending birthday cards to the employees is a positive practice, but it would not be as effective as direct communication that could take place face-to-face.

26. Service Consultant A always gives the highest priority to answering the phone rather than dealing with customers in the write-up area. Service Consultant B is able to carry on a phone conversation while dealing with a customer in person. Who is correct?

TASK A.1.1

    A. A only
    B. B only
    C. Both A and B
    D. Neither A nor B

    Answer A is incorrect. A service consultant should be a well-balanced, courteous professional in how she handles customers. It is not wise to always give priority to the phone, because the customers in the write-up area will likely be offended if they do not get waited on in a timely manner.

    Answer B is incorrect. It is not advisable to try to do too many things at once. In the example, the service consultant will likely offend both customers who are being dealt with. It is wise to work with each customer as quickly and effectively as possible and then move to the next customer.

    Answer C is incorrect. Neither Service Consultant is correct.

    Answer D is correct. Neither Service Consultant is correct. It is wise to attend to the customers (in person and on the phone) in the order that they present themselves. Giving all of your attention to either one while ignoring the other is not going to be very effective.

27. Service Consultant A says that the crankshaft sensor sends engine speed data to the engine computer. Service Consultant B says that the crankshaft sensor needs to be mounted near a reluctor ring. Who is correct?

TASK B.1.1

    A. A only
    B. B only
    C. Both A and B
    D. Neither A nor B

    Answer A is incorrect. Service Consultant B is also correct.

    Answer B is incorrect. Service Consultant A is also correct.

    Answer C is correct. Both Service Consultants are correct. The crankshaft sensor is mounted near a reluctor ring on the crankshaft. It sends a speed signal to the engine computer that is vital to engine operation.

    Answer D is incorrect. Both Service Consultants are correct.

TASK B.1.1

28. Service Consultant A says that the alternator needs to be mounted near the flywheel. Service Consultant B says that the starter is driven by the accessory drive belt. Who is correct?

 A. A only
 B. B only
 C. Both A and B
 D. Neither A nor B

Answer A is incorrect. The alternator is mounted on the front of the engine and is driven by a belt that connects to the crankshaft.

Answer B is incorrect. The starter receives an electrical signal from the ignition switch and creates motion that engages the engine to start.

Answer C is incorrect. Neither Service Consultant is correct.

Answer D is correct. Neither Service Consultant is correct. The alternator is mounted on the front of the engine and is driven by a belt. The starter engages the flywheel to start the engine after receiving a signal from the ignition switch.

TASK B.1.2

29. Which component is LEAST LIKELY to be needed in an overhead cam engine design?

 A. Pushrod
 B. Rocker arm
 C. Intake valve
 D. Exhaust valve

Answer A is correct. Overhead cam engines do not use pushrods. Typically, the camshaft engages some type of rocker assembly to open the valves.

Answer B is incorrect. Overhead cam engine still need rocker arms to transfer motion to the valves.

Answer C is incorrect. Overhead cam engines still need intake valves to control the flow of fuel and air into the engine.

Answer D is incorrect. Overhead cam engines still need exhaust valves to control the flow of burned gases out of the engine.

TASK A.1.6

30. Service Consultant A promotes the repair shop by being knowledgeable about the certification status of the technicians who are employed there. Service Consultant B promotes the repair shop by being knowledgeable about the professionalism of the technicians who are employed there. Who is correct?

 A. A only
 B. B only
 C. Both A and B
 D. Neither A nor B

Answer A is incorrect. Service Consultant B is also correct.

Answer B is incorrect. Service Consultant A is also correct.

Answer C is correct. Both Service Consultants are correct. Professional service consultants should actively promote the repair shop by pointing out the positive attributes of the service technicians who work there.

Answer D is incorrect. Both Service Consultants are correct.

31. Which component would be most likely to be replaced during an intake manifold gasket replacement?

TASK B.1.3

A. Rod bearing
B. Main bearing
C. Thermostat
D. Oil pump

Answer A is incorrect. A rod bearing is not a related component that would be replaced during an intake manifold gasket replacement.

Answer B is incorrect. A main bearing is not a related component that would be replaced during an intake manifold gasket replacement.

Answer C is correct. The thermostat is typically mounted on or near the intake manifold and could be easily serviced during an intake manifold gasket replacement.

Answer D is incorrect. The oil pump is not typically replaced during an intake manifold gasket replacement. The oil pump is located near the front timing cover or in the oil pan area of the engine.

32. Service Consultant A says that the clutch disc is located between the pressure plate and the flywheel. Service Consultant B says that the throw out bearing is located the end of the crankshaft. Who is correct?

TASK B.2.1

A. A only
B. B only
C. Both A and B
D. Neither A nor B

Answer A is correct. Only Service Consultant A is correct. The clutch disc is located between the pressure plate and the flywheel. This component allows the engine to be disconnected from the manual transmission when the clutch pedal is pressed.

Answer B is incorrect. The throw out bearing is located on a stationary shaft near the front of the manual transmission. This component engages the pressure plate when the clutch pedal is depressed.

Answer C is incorrect. Only Service Consultant A is correct.

Answer D is incorrect. Service Consultant A is correct.

33. Which drive train component connects the transaxle to the drive hubs?

TASK B.2.1, B.2.2

A. Hub bearing
B. Axle half shaft
C. Universal joint
D. Final drive assembly

Answer A is incorrect. The hub bearing is used on most late-model vehicles to minimize rolling friction of the wheel.

Answer B is correct. The axle half shaft connects the transaxle to the drive hubs on front-wheel drive vehicles. This component typically has constant velocity (CV) joints on each end to allow the drive angle to change as the suspension moves up and down.

Answer C is incorrect. A universal joint is used on driveshafts of rear-wheel drive vehicles. This component allows the drive angle to change as the rear suspension moves up and down.

Answer D is incorrect. A final drive assembly is typically a part of a front-wheel drive transaxle assembly.

TASK A.1.13

34. A follow-up call is being completed. Service Consultant A asks the customer if he/she is pleased with the cleanliness of the repair shop. Service Consultant B asks the customer if he/she was pleased with the location of the shop. Who is correct?

A. A only
B. B only
C. Both A and B
D. Neither A nor B

Answer A is correct. Only Service Consultant A is correct. Asking a customer about the cleanliness of the shop is a likely question to ask on a follow-up call. It is important to give good impressions to the customer; having a clean and neat shop goes a long way in making a good impression.

Answer B is incorrect. The shop location is not something that would be inquired about on a follow-up call. The repair shop is not going to be able to change the location of the shop.

Answer C is incorrect. Only Service Consultant A is correct.

Answer D is incorrect. Service Consultant A is correct.

TASK B.3.1

35. All of the following components are parts of the steering system EXCEPT:

A. Rack and pinion gear
B. Strut
C. Steering shaft
D. Tie rod end

Answer A is incorrect. The rack and pinion gear is a common component of steering systems on late-model vehicles. This component transfers rotational motion into lateral motion, which moves the front wheels side to side.

Answer B is correct. The strut is part of the suspension system. This device helps dampen the compressions of the vehicle suspension system.

Answer C is incorrect. The steering shaft is a common component of the steering system. This component connects the steering wheel to the steering gear.

Answer D is incorrect. The tie rod end is a common component of the steering system. This component connects the steering gear to the steering knuckle.

TASK B.3.2

36. Which alignment angle is described as the inward- or outward-rolling direction of the tires?

A. Toe
B. Camber
C. Caster
D. Thrust angle

Answer A is correct. The inward- and outward-rolling direction of the tires is called toe. If the wheels are pointed outward, it is described as toe-out. If the wheels are pointed inward, it is called toe-in.

Answer B is incorrect. Camber is the angle of the tires as they lean in or out at the top of the tire. Positive camber is when the tires lean out beyond a straight vertical line. Negative camber is when the tires lean inward at the top of the tire.

Answer C is incorrect. Caster is the alignment angle that is created by the upper and lower ball joints.

Answer D is incorrect. Thrust angle is the overall angle of the rear toe setting in relation to the center line of the vehicle.

37. Which of the following antilock brake components is most likely to be located near each wheel?

TASK B.3.3

A. Modulator assembly
B. Speed sensor
C. Electronic control unit
D. Deceleration sensor

Answer A is incorrect. The antilock brake modulator is usually located in the engine compartment area.

Answer B is correct. The speed sensor is typically mounted near each wheel in order to sense the speed of each wheel. Some speed sensors are built into the hub assembly and some sensors are separate from the hub.

Answer C is incorrect. The electronic control unit is typically mounted in the engine compartment or in the cab area.

Answer D is incorrect. The deceleration sensor is typically mounted in the cab area on the floor pan of the body.

38. Service Consultant A says that improved fuel economy would be a benefit of having well-maintained coolant. Service Consultant B says that improved vehicle handling would be a benefit of having a four-wheel alignment performed. Who is correct?

TASK A.2.5

A. A only
B. B only
C. Both A and B
D. Neither A nor B

Answer A is incorrect. Servicing the coolant on a regular basis is a good practice, but would not likely have a direct impact on fuel economy.

Answer B is correct. Only Service Consultant B is correct. Four-wheel alignments contribute to improved vehicle handling.

Answer C is incorrect. Only Service Consultant B is correct.

Answer D is incorrect. Service Consultant B is correct.

39. The HVAC system performs all of the following functions for a vehicle EXCEPT:

TASK B.4.2

A. Humidifies the air in dry conditions
B. Heats the cabin when needed
C. Cools the cabin when needed
D. Dehumidifies the air when the A/C compressor is operated

Answer A is correct. The HVAC system does not add moisture to the air at any time.

Answer B is incorrect. The HVAC system heats the air by routing it through the heater core when the controls are set to a warm/hot setting.

Answer C is incorrect. The HVAC system cools air by routing the air through the evaporator core and by-passing the heater core when the controls are set to the cool setting.

Answer D is incorrect. The HVAC system dehumidifies the air when the A/C compressor is operated. The compressor typically operates in the A/C mode as well as the Defrost mode.

TASK A.2.1, B.6.1, B.6.4

40. A customer who has a failed engine that was obviously caused from the engine oil never being changed does not understand why his/her vehicle cannot be repaired under warranty. Service Consultant A is very harsh in explaining that the customer should have maintained the vehicle by having the oil changed on a regular basis. Service Consultant B explains the situation to the customer by showing him/her how deteriorated the oil was. Who is correct?

A. A only
B. B only
C. Both A and B
D. Neither A nor B

Answer A is incorrect. The service consultant should never be harsh when dealing with an upset customer. It is wise to use calm, logical explanations when handling these situations.

Answer B is correct. Only Service Consultant B is correct. It is important to use a calm demeanor when discussing vehicle repairs with customers. Using logical facts and educating the customer is the best way to handle stressful situations.

Answer C is incorrect. Only Service Consultant B is correct.

Answer D is incorrect. Service Consultant B is correct.

TASK B.1.1, B.5.1

41. Service Consultant A says that the throttle plates should be cleaned during a throttle body service. Service Consultant B says that some throttle bodies have to be removed from the engine to be cleaned. Who is correct?

A. A only
B. B only
C. Both A and B
D. Neither A nor B

Answer A is incorrect. Service Consultant B is also correct.

Answer B is incorrect. Service Consultant A is also correct.

Answer C is correct. Both Service Consultants are correct. The throttle plates are cleaned during a throttle body service to allow for better airflow into the engine. Some throttle bodies have to be removed from the engine in order to be cleaned.

Answer D is incorrect. Both Service Consultants are correct.

TASK A.1.11

42. A customer arrives at the repair shop to pick up a fleet vehicle after the shop has performed a warranty repair. Service Consultant A inquires about how many vehicles are in the fleet. Service Consultant B gives the customer a card and requests that the fleet manager call at the nearest convenient time to discuss future business. Who is correct?

A. A only
B. B only
C. Both A and B
D. Neither A nor B

Answer A is incorrect. Service Consultant B is also correct.

Answer B is incorrect. Service Consultant A is also correct.

Answer C is correct. Both Service Consultants are correct. Service consultants should always be looking for opportunities to increase the business of the repair shop. Fleet customers are very lucrative accounts for a repair shop to acquire.

Answer D is incorrect. Both Service Consultants are correct.

43. Which is the LEAST LIKELY place to find a maintenance schedule for a vehicle?

TASK B.5.3

   A. Owner's manual
   B. Warranty booklet
   C. Electronic database
   D. Shop manual

Answer A is incorrect. Most owner's manuals would have a maintenance schedule for the vehicle.

Answer B is correct. The warranty booklet would not likely contain the maintenance schedule.

Answer C is incorrect. An electronic database would likely have maintenance schedules for most vehicles.

Answer D is incorrect. The shop manual would likely have a maintenance schedule for the vehicle.

44. Which of the following examples describes how each customer should be greeted as he/she arrives at the repair facility?

TASK A.1.4

   A. Without emotion or any facial expression
   B. With a slap on the back and an off-color joke
   C. With a smile and a cordial greeting
   D. By sharing with them your daily problems in the service department

Answer A is incorrect. A customer will pick up on body language and facial expression quickly and feel uncomfortable if they sense negativity in the service consultant.

Answer B is incorrect. A cordial greeting is appropriate, but telling jokes that could offend others in the area is not advisable.

Answer C is correct. Customers appreciate being greeted with a smile and a cordial greeting.

Answer D is incorrect. It is not advisable to share any negative issues that may be going on in your life or in the service department.

45. Service Consultant A says that items such as belts and hoses are typically considered wear items and are not typically covered by extended warranties. Service Consultant B says that major transmission components are considered power train items and would usually be covered by extended warranties. Who is correct?

TASK B.6.1, B.6.3

   A. A only
   B. B only
   C. Both A and B
   D. Neither A nor B

Answer A is incorrect. Service Consultant B is also correct.

Answer B is incorrect. Service Consultant A is also correct.

Answer C is correct. Both Service Consultants are correct. Normal wear items are not usually covered by extended warranties, but major power train items would be covered by these warranties.

Answer D is incorrect. Both Service Consultants are correct.

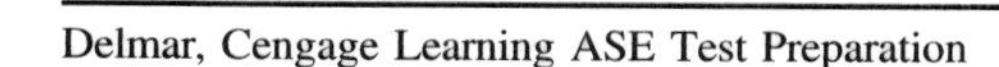

TASK B.7.1

46. Service Consultant A says that the tenth digit of the vehicle identification number (VIN) represents the engine. Service Consultant B says that the eighth digit of the VIN is the country of origin. Who is correct?

A. A only
B. B only
C. Both A and B
D. Neither A nor B

Answer A is incorrect. The tenth digit of the VIN represents the model year.

Answer B is incorrect. The eighth digit of the VIN represents the engine.

Answer C is incorrect. Neither Service Consultant is correct.

Answer D is correct. Neither Service Consultant is correct. The tenth digit of the VIN is the model year and the eighth digit of the VIN is the engine. The first digit of the VIN is the country of origin.

TASK B.7.4

47. Service Consultant A says that a sedan is a car that has a retractable hardtop. Service Consultant B says that a coupe is a car that has four doors. Who is correct?

A. A only
B. B only
C. Both A and B
D. Neither A nor B

Answer A is incorrect. A sedan is a vehicle that has four doors.

Answer B is incorrect. A coupe is a vehicle that has two doors.

Answer C is incorrect. Neither Service Consultant is correct.

Answer D is correct. Neither Service Consultant is correct. A sedan is a vehicle that has four doors and a coupe is a vehicle that has two doors.

TASK A.1.8

48. A customer is inquiring how often the tires should be rotated. Service Consultant A says the tires need to be rotated each time the oil is changed. Service Consultant B says the tires should be rotated at each 30,000-mile service. Who is correct?

A. A only
B. B only
C. Both A and B
D. Neither A nor B

Answer A is incorrect. Oil change intervals can vary depending on recommendations from various manufacturers, as well as the type of oil being used, thus possibly making the rotation interval too long or too short.

Answer B is incorrect. 30,000 miles is too long to go between tire rotations.

Answer C is incorrect. Neither Service Consultant is correct.

Answer D is correct. Neither Service Consultant is correct. The tires should be rotated according to the tire manufacturer recommendations. Many intervals are at about 7,500 miles.

49. Service Consultant A says that some repair shops choose to sublet radiator repairs to an outside facility. Service Consultant B says that some repair shops choose to sublet oil changes to an outside facility. Who is correct?

TASK C.2

A. A only
B. B only
C. Both A and B
D. Neither A nor B

Answer A is correct. Only Service Consultant A is correct. Some shops choose to sublet radiator repairs to facilities that specialize in performing this service.

Answer B is incorrect. It is very unlikely that an auto repair shop would sublet an oil change to an outside facility. This service builds trust for a repair shop and also offers an opportunity for evaluating other needed services while the oil is being changed.

Answer C is incorrect. Only Service Consultant A is correct.

Answer D is incorrect. Service Consultant A is correct.

50. A customer calls and says that he nearly had an accident due to the engine stalling while crossing an intersection. The shop replaced a fuel pump three days ago. Service Consultant A recommends that the customer return to the service facility the next day to recheck the fuel system. Service Consultant B offers to send a wrecker to pick up the vehicle immediately, due to the potentially unsafe conditions of the vehicle. Who is correct?

TASK C.4

A. A only
B. B only
C. Both A and B
D. Neither A nor B

Answer A is incorrect. The Service Consultant should not allow the vehicle to be driven any more before it is checked out thoroughly.

Answer B is correct. Only Service Consultant B is correct. Sending a wrecker is the right reaction to this situation, due to the serious safety liability involved.

Answer C is incorrect. Only Service Consultant B is correct.

Answer D is incorrect. Service Consultant B is correct.

PREPARATION EXAM 5—ANSWER KEY

1. A
2. A
3. D
4. A
5. D
6. B
7. A
8. B
9. B
10. A
11. A
12. A
13. A
14. A
15. B
16. A
17. C
18. B
19. C
20. A
21. A
22. D
23. C
24. C
25. C
26. A
27. D
28. D
29. D
30. C
31. A
32. D
33. A
34. B
35. D
36. B
37. B
38. C
39. B
40. C
41. D
42. A
43. C
44. B
45. C
46. D
47. C
48. C
49. C
50. B

PREPARATION EXAM 5—EXPLANATIONS

TASK A.1.1

1. A customer calls to speak with a service consultant who is already working with a customer. What should the service consultant taking the call do?

A. Take the customer's name and number and assure them they will be called back.

B. Transfer the customer to the parts department.

C. Place the customer on hold until the consultant is available.

D. Transfer the call to the service manager.

Answer A is correct. This is very important to customers. If you take a message, be sure you get it to the person for whom it is intended. This is also the best solution to avoid incomplete and rushed customer service.

Answer B is incorrect. The parts department will not be of any help to the customer.

Answer C is incorrect. By placing the customer on hold, you are putting pressure on the other consultant to finish with the customer with whom they are already working. This effectively irritates two customers and may not leave time for a complete close to the consultant's current customer interaction.

Answer D is incorrect. The service manager is not going to respond to the calling customer's issue. The service manager should be the last link in the customer service chain.

2. Technician efficiency needs to be monitored continuously. Service Consultant A monitors the weekly flat rate hours that each technician completes. Service Consultant B times each technician on every job and compares this time with the flat rate time. Who is correct?

TASK A.3.7

   A. A only
   B. B only
   C. Both A and B
   D. Neither A nor B

   Answer A is correct. Only Service Consultant A is correct. Tracking the flat rate hours that each technician completes is a good indicator of the efficiency for the technician. The production for technicians will vary from week to week, so an average should be gathered over a several week period to get a good indicator of efficiency.

   Answer B is incorrect. The Service Consultant will not have time to record the time on every job that the technicians complete. It would be better to look at larger blocks of time to monitor efficiency.

   Answer C is incorrect. Only Service Consultant A is correct.

   Answer D is incorrect. Service Consultant A is correct.

3. Which of the following terms is defined as a manufacturer's technical document created to assist technicians in repairing pattern failures more quickly?

TASK B.6.3

   A. Service contract
   B. Campaign/recall
   C. Warranty
   D. Technical service bulletin (TSB)

   Answer A is incorrect. A service contract is a type of extended warranty on some vehicles. Service consultants need to be proficient in dealing with types of insurance plans. Detailed and accurate estimates are required before calling these companies for approval.

   Answer B is incorrect. A campaign/recall is a program that manufacturers create to invite customers to bring their vehicles back in for a free repair in order to correct a safety fault in the vehicle.

   Answer C is incorrect. A warranty is a type of insurance plan that comes with all new cars and can be for various lengths of time and/or mileage. Many used vehicles can still have the remainder of the factory warranty in place. Filing the paperwork correctly is a requirement in filing warranty claims so that the service facility is properly compensated.

   Answer D is correct. A TSB is written by manufacturers to assist the technician with repairing the vehicle the first time.

4. All of the following are appropriate greeting approaches for a service consultant EXCEPT:

TASK A.1.4

   A. Warm hug
   B. Friendly demeanor
   C. Handshake
   D. Smile

   Answer A is correct. Service consultants do not offer this level of greeting to their repair shop customers.

   Answer B is incorrect. Service consultants should try to have a friendly demeanor when greeting customers. This characteristic helps the customers feel welcome in the service facility.

   Answer C is incorrect. Offering a handshake is an appropriate greeting for a service consultant to practice for customers entering the service shop.

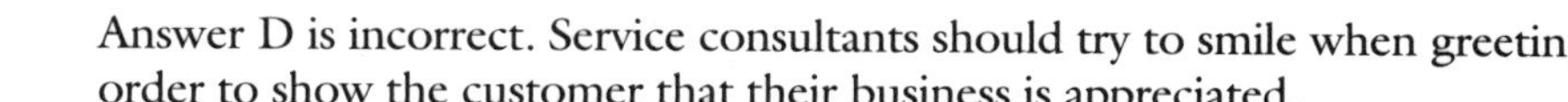

   Answer D is incorrect. Service consultants should try to smile when greeting customers in order to show the customer that their business is appreciated.

TASK B.1.3

5. Which of the following components is integrated into the exhaust system?

   A. Knock sensor
   B. Camshaft sensor
   C. Crankshaft sensor
   D. Oxygen sensor

Answer A is incorrect. The knock sensor is mounted on the engine. It signals the engine computer when detonation occurs in the engine.

Answer B is incorrect. The camshaft sensor is mounted on the engine. It signals the engine computer as the camshaft rotates.

Answer C is incorrect. The crankshaft sensor is mounted on the engine. It signals the engine computer as the crankshaft rotates.

Answer D is correct. The oxygen sensor is mounted in the exhaust system. It signals the engine computer as the oxygen content of the exhaust changes.

TASK A.1.6, A.1.7

6. A customer's vehicle is being serviced for the first time by your shop. The customer states that she has had the same problem worked on at two other shops. Service Consultant A says that the technicians at this shop are much more competent than the other shops. Service Consultant B asks the customer to bring her repair records along to help the shop get an idea of the vehicle's history. Who is correct?

   A. A only
   B. B only
   C. Both A and B
   D. Neither A nor B

Answer A is incorrect. A service consultant should refrain from expressing a negative judgment about other shops in front of customers. The customers can make up their own minds about other shops on their own.

Answer B is correct. Only Service Consultant B is correct. Having the records from previous repair attempts will assist the shop in diagnosing the vehicle correctly.

Answer C is incorrect. Only Service Consultant B is correct.

Answer D is incorrect. Service Consultant B is correct.

TASK C.1

7. A customer comes in at 3 p.m. needing a complete brake job performed. The estimated time for this repair is four hours and the shop closes at 5 p.m. Service Consultant A recommends that the customer drop the car off and plan on picking the car up the next day at noon. Service Consultant B says that the shop can complete the job that day and encourages the technician to rush to complete the job on time. Who is correct?

   A. A only
   B. B only
   C. Both A and B
   D. Neither A nor B

Answer A is correct. Only Service Consultant A is correct. An efficient service consultant will sell the job, but not create unrealistic expectations for the customer.

Answer B is incorrect. It is not advisable to encourage the technician to rush a vehicle repair due to implications of reduced quality control and reduced vehicle safety.

Answer C is incorrect. Only Service Consultant A is correct.

Answer D is incorrect. Service Consultant A is correct.

8. A service consultant has prepared an estimate based upon a technician's diagnosis. Which of the following should the service consultant do first before providing the customer with the estimate?

TASK A.1.9, A.3.3

   A. Check with the technician to plan a completion time.
   B. Check the availability of the repair parts.
   C. Test drive the vehicle.
   D. Identify additional needed services.

   Answer A is incorrect. The technician does not typically keep up with the work flow in the shop, so he will not know when to promise the work to be completed.

   Answer B is correct. The other items should be done at various times during the repair process, but before you decide on a completion time to tell the customer, check to see when you can have the parts in your hands. How much and how long are the most frequently asked questions by customers.

   Answer C is incorrect. A test drive might be performed by either the service consultant or the technician prior to preparing the estimate.

   Answer D is incorrect. Identifying additional needed services is not the first step in completing the sale.

9. Service Consultant A says that the tire pressure warning light must be reset at each oil and filter change service. Service Consultant B says that the tire pressure warning light will sometimes illuminate as the weather gets colder. Who is correct?

TASK B.5.2

   A. A only
   B. B only
   C. Both A and B
   D. Neither A nor B

   Answer A is incorrect. The tire pressure warning light does not have to be reset at each oil change. This system monitors tire pressure in each tire and illuminates a light if pressure drops below specifications.

   Answer B is correct. Only Service Consultant B is correct. It is common for the air pressure to decrease in colder temperatures, which could cause the tire pressure warning light to illuminate.

   Answer C is incorrect. Only Service Consultant B is correct.

   Answer D is incorrect. Service Consultant B is correct.

10. A customer calls and states that his/her vehicle has an electrical problem that has been recurring after three attempts to repair it. Which of these should the service consultant do first?

TASK A.1.1, A.1.11

   A. Check the repair history to see if this shop has worked on this vehicle's electrical problem before.
   B. Offer to diagnose the vehicle free of charge.

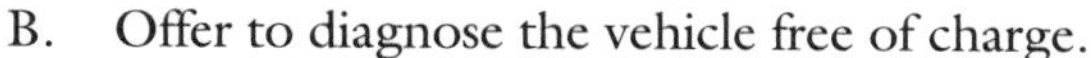

   C. Ask the customer to explain in detail all that has been done to the vehicle.
   D. Explain that some problems require several attempts to fix.

   Answer A is correct. A Service Advisor should determine if repairs are comeback/repeat repairs during the initial communication with the customer, if possible.

   Answer B is incorrect. A service consultant should never volunteer to give a diagnosis without charging for it during the initial phone conversation.

   Answer C is incorrect. A Service Advisor should never ask for a long or detailed explanation on the phone because the customer will not be able to accurately remember each detail. It is a good idea to have them bring any service records from other repair shops with them when they come to this repair shop.

   Answer D is incorrect. A Service Advisor should never state that repeated repair attempts are normal practice.

TASK B.2.2

11. Service Consultant A says that the automatic transmission fluid should be changed more frequently if the vehicle is used to pull trailers with heavy loads. Service Consultant B says that all transmission fluid is the same as long as a name brand of fluid is used. Who is correct?

    A. A only
    B. B only
    C. Both A and B
    D. Neither A nor B

    Answer A is correct. Only Service Consultant A is correct. Pulling a load on a regular basis causes the transmission fluid to run at a hotter temperature, which causes it to break down sooner. The fluid should be changed more frequently on vehicles that pull loads regularly.

    Answer B is incorrect. There are several different types of transmission fluid available and care should be taken to install the correct fluid for each vehicle that is serviced.

    Answer C is incorrect. Only Service Consultant A is correct.

    Answer D is incorrect. Service Consultant A is correct.

TASK A.1.14

12. A customer is picking up his/her vehicle from a repair shop. Service Consultant A says that it is important to take the time to explain the work performed in as much detail as the customer requires. Service Consultant B says that if the customer asks questions it indicates he/she does not trust the shop. Who is correct?

    A. A only
    B. B only
    C. Both A and B
    D. Neither A nor B

    Answer A is correct. Only Service Consultant A is correct. Different customers require different levels of explanation. If a customer wants lots of details, that does not automatically mean he/she distrusts you. It usually means the customer takes a very active part in maintaining the vehicle and wants to understand the process. These can be your very best customers. This is a great chance for you to show off your expertise and product and repair knowledge.

    Answer B is incorrect. Service consultants should expect their customers to ask questions about the repair of their vehicle.

    Answer C is incorrect. Only Service Consultant A is correct.

    Answer D is incorrect. Service Consultant A is correct.

TASK B.4.3

13. Service Consultant A says that the starter relay is part of the starting system. Service Consultant B says that the voltage regulator is part of the starting system. Who is correct?

    A. A only
    B. B only
    C. Both A and B
    D. Neither A nor B

    Answer A is correct. Only Service Consultant A is correct. The starter relay is part of the starting system. This component sends a voltage signal down to the starter motor when the driver activates the start/ignition switch.

    Answer B is incorrect. The voltage regulator is part of the charging system. This component controls the output of the alternator as the electrical load varies.

    Answer C is incorrect. Only Service Consultant A is correct.

    Answer D is incorrect. Service Consultant A is correct.

14. A vehicle is in the repair shop for an inspection. The technician recommends having the throttle body cleaned for a stalling problem and the valve cover gasket replaced for a small oil leak. The customer asks the service consultant to prioritize the two services in order of importance. Service Consultant A recommends the throttle body cleaning because the stalling problem is dangerous. Service Consultant B says that both items are of equal importance. Who is correct?

TASK A.2.2

 A. A only
 B. B only
 C. Both A and B
 D. Neither A nor B

Answer A is correct. Only Service Consultant A is correct. The dirty throttle body is a safety issue due to the potential stalling problem. This problem should be repaired very soon.

Answer B is incorrect. A small oil leak at the valve cover gasket is not considered a safety-related problem. It is not as high a priority as the dirty throttle body.

Answer C is incorrect. Only Service Consultant A is correct.

Answer D is incorrect. Service Consultant A is correct.

15. A technician turns in a repair order that recommends replacement of the constant velocity (CV) boot with no further description. Service Consultant A calls the customer to get approval to complete the repair. Service Consultant B consults with the technician to find out the reason for replacement of the boot. Who is correct?

TASK A.1.7, A.1.8, A.3.2

 A. A only
 B. B only
 C. Both A and B
 D. Neither A nor B

Answer A is incorrect. The service consultant needs to find out the problem with the CV boot prior to calling the customer. The customer will likely want to know why the part is being recommended.

Answer B is correct. Only Service Consultant B is correct. The service consultant should get an explanation of why the CV boot needs replacing prior to checking with the customer for approval. The customer will likely want to know why the part is being recommended.

Answer C is incorrect. Only Service Consultant B is correct.

Answer D is incorrect. Service Consultant B is correct.

16. Service Consultant A says the benefit of a cabin air filter is that it reduces contaminants in the cabin area. Service Consultant B says that a benefit of a cabin air filter is cooler air from the vents. Who is correct?

TASK A.2.5, B.4.2

 A. A only
 B. B only
 C. Both A and B
 D. Neither A nor B

Answer A is correct. Only Service Consultant A is correct. The cabin air filter is used in the HVAC system to reduce the air contaminants in the cabin area. These filters should be serviced periodically according to manufacturer recommendations.

Answer B is incorrect. The cabin air filter does not improve the cooling capacity of the HVAC system.

Answer C is incorrect. Only Service Consultant A is correct.

Answer D is incorrect. Service Consultant A is correct.

TASK B.7.4

17. Service Consultant A says that vehicles that do not have frames around the windows are known as hard tops. Service Consultant B says that a hatchback is a vehicle with a rear trunk/window combination that lifts up. Who is correct?

    A. A only
    B. B only
    C. Both A and B
    D. Neither A nor B

Answer A is incorrect. Service Consultant B is also correct.

Answer B is incorrect. Service Consultant A is also correct.

Answer C correct. Both Service Consultants are correct. There are many body styles. Since there can be many body styles in a single model of vehicle, it is a good idea to be able to identify them visually. Remember, if the door glass closes directly against a weather strip, the vehicle is a hard top, unless it is a convertible. Body styles can be mixed together, as in the 5-door hatchback that is both a sedan and a hatchback.

Answer D is incorrect. Both Service Consultants are correct.

TASK A.2.1

18. Service Consultant A says that providing an estimate is required by law in all states. Service Consultant B says that explaining the details of the estimate will give customers a better understanding of what is going to be done to their vehicle. Who is correct?

    A. A only
    B. B only
    C. Both A and B
    D. Neither A nor B

Answer A is incorrect. Not all states require that service repair shops provide an estimate.

Answer B is correct. Only Service Consultant B is correct. The Service Consultant needs to logically explain the details of the estimate to customers in order to gain their confidence and trust.

Answer C is incorrect. Only Service Consultant B is correct.

Answer D is incorrect. Service Consultant B is correct.

TASK A.1.14, A.2.3

19. Service Consultant A provides clear and understandable answers to detailed questions that customers have. Service Consultant B shows customers the details of the repair order before completing the repair order. Who is correct?

    A. A only
    B. B only
    C. Both A and B
    D. Neither A nor B

Answer A is incorrect. Service Consultant B is also correct.

Answer B is incorrect. Service Consultant A is also correct.

Answer C is correct. Both Service Consultants are correct. The service consultant should be prepared to disclose all pertinent information to the customer during the repair process. In addition, the service consultant should take the time to provide clear answers to any questions that the customer might have.

Answer D is incorrect. Both Service Consultants are correct.

20. A vehicle is in the repair shop for a water pump replacement. Service Consultant A recommends that the cooling system be totally flushed and refilled with new coolant while the water pump is being replaced. Service Consultant B recommends that the A/C system should be recovered and recharged while the water pump is being replaced. Who is correct?

TASK A.2.4, B.1.2, B.4.3

A. A only
B. B only
C. Both A and B
D. Neither A nor B

Answer A is correct. Only Service Consultant A is correct. Servicing the cooling system would be highly advisable to perform at the time of a water pump replacement. The coolant would be drained during the process of replacing the water pump.

Answer B is incorrect. Servicing the A/C system would not be considered a related repair to the water pump replacement.

Answer C is incorrect. Only Service Consultant A is correct.

Answer D is incorrect. Service Consultant A is correct.

21. Service Consultant A says that a benefit to having regular brake inspections is reliable stopping operation of the vehicle. Service Consultant B says that a benefit to regular brake inspections is improved fuel economy. Who is correct?

TASK A.2.5

A. A only
B. B only
C. Both A and B
D. Neither A nor B

Answer A is correct. Only Service Consultant A is correct. Regular brake inspections would provide reliable stopping operation of the vehicle.

Answer B is incorrect. Fuel economy would not likely be affected by performing regular brake inspections.

Answer C is incorrect. Only Service Consultant A is correct.

Answer D is incorrect. Service Consultant A is correct.

22. Service Consultant A says that an accurate estimate is NOT critical to have before calling the customer for approval. Service Consultant B says that an exact completion time is important to have before calling for customer approval. Who is correct?

TASK A.1.10, A.2.6

A. A only
B. B only
C. Both A and B
D. Neither A nor B

Answer A is incorrect. The service consultant does need to have an accurate estimate prior to calling the customer for approval of the repair.

Answer B is incorrect. The service consultant does not need an exact completion time calculated before calling the customer for approval. However, an estimated completion time would be necessary to have before notifying the customer.

Answer C is incorrect. Neither Service Consultant is correct.

Answer D is correct. Neither Service Consultant is correct.

TASK A.3.1

23. A customer has just given approval for repair of his/her vehicle. Service Consultant A says the technician should be provided with the approved work order. Service Consultant B says documentation of the customer's approval should be on the work order. Who is correct?

A. A only
B. B only
C. Both A and B
D. Neither A nor B

Answer A is incorrect. Service Consultant B is also correct.

Answer B is incorrect. Service Consultant A is also correct.

Answer C is correct. Both Service Consultants are correct. Once the customer has approved a repair, the service consultant should note the approval details on the repair order and then give the repair order to the technician.

Answer D is incorrect. Both Service Consultants are correct.

TASK A.2.4

24. Recommending additional services is a common practice for service consultants. Service Consultant A says that the shop will be more profitable due to increased labor charges if additional services are sold. Service Consultant B says that customers will appreciate the fact that the technicians are inspecting their vehicles for safety and maintenance items. Who is correct?

A. A only
B. B only
C. Both A and B
D. Neither A nor B

Answer A is incorrect. Service Consultant B is also correct.

Answer B is incorrect. Service Consultant A is also correct.

Answer C is correct. Both Service Consultants are correct. Service consultants can increase the profitability of a shop, in addition to building the confidence of customers, by recommending additional services. This practice should be handled with care to avoid the appearance that the shop is taking advantage of the customer by recommending services without merit.

Answer D is incorrect. Both Service Consultants are correct.

TASK A.3.5

25. Which is the LEAST LIKELY method to be used when checking for quality control in a repair shop?

A. Keep a log of repeat repairs for each technician.
B. Keep a total of labor "write-off" amounts for each technician.
C. Keep a log of redeemed coupons for each month.
D. Keep a log of repeat repairs for the whole shop.

Answer A is incorrect. Keeping a log of repeat repairs for each technician would be a way to monitor the quality of the repairs being made by the shop.

Answer B is incorrect. Monitoring the "write-off" labor amounts is a way to track the quality control for a shop.

Answer C is correct. Keeping up with the redeemed coupons would not give any indication about quality control in the shop. This data would only give feedback about the marketing strategies that the shop is using.

Answer D is incorrect. Keeping a log of repeat repairs for the whole shop is a way to monitor the quality control for the shop.

26. A service consultant has just completed compiling and writing up a customer's concerns. Which of the following should she do next?

A. Confirm the accuracy of the information on the repair order.
B. Arrange a ride home for the customer.
C. Offer an estimate for the repairs.
D. Have the porter wash the car.

Answer A is correct. The service consultant should always confirm the accuracy of the repair order prior to letting the customer leave the write-up area.

Answer B is incorrect. Arranging transportation for the customer would be done after confirming the accuracy of the repair order.

Answer C is incorrect. An estimate is not typically provided until the vehicle is diagnosed.

Answer D is incorrect. The car would not be washed until the repairs have been completed.

27. Which engine component regulates engine temperature?

A. Heater core
B. Radiator
C. Water pump
D. Thermostat

Answer A is incorrect. The heater core is a heat exchanger in the HVAC duct that delivers heat to the cabin when the duct air is routed through it.

Answer B is incorrect. The radiator is a heat exchanger located at the front of the vehicle that releases the heat from the coolant to the ambient air.

Answer C is incorrect. The water pump causes the coolant to be moved throughout the engine and is typically driven by a belt or chain.

Answer D is correct. The engine thermostat controls the flow of coolant to the radiator to regulate the engine temperature.

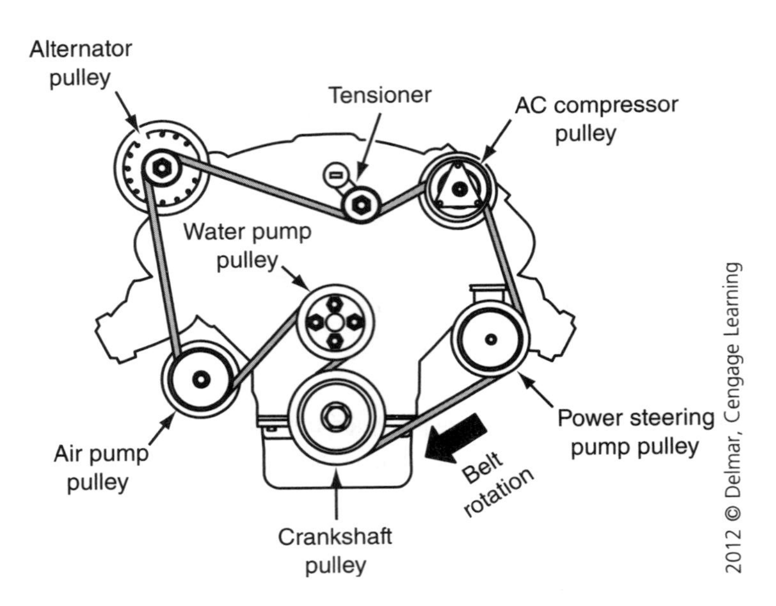

TASK B.1.2

28. Referring to the figure above, Service Consultant A says that the alternator provides the belt tension on the drive belt. Service Consultant B says that the belt used in the figure is a V-belt. Who is correct?

A. A only
B. B only
C. Both A and B
D. Neither A nor B

Answer A is incorrect. The belt tensioner provides the tension on the serpentine belt.

Answer B is incorrect. The belt used in the figure is a serpentine belt since it is used in conjunction with a belt tensioner. Vehicles that use V-belts do not use an automatic tensioner.

Answer C is incorrect. Neither Service Consultant is correct.

Answer D is correct. Neither Service Consultant is correct. The alternator does not provide the belt tension on the belt in the figure. The serpentine belt is held tight by the tensioner. The tensioner is a spring-loaded component that constantly loads the belt with pressure.

TASK B.1.2

29. Which of the following components is part of a typical charging system?

A. Starter solenoid
B. Park/neutral switch
C. Ignition switch
D. Voltage regulator

Answer A is incorrect. The starter solenoid is part of the starting system. This component provides the starter with a high-current path when the start switch is activated.

Answer B is incorrect. The park/neutral switch is part of the starting system. This component prevents the vehicle from starting while in gear.

Answer C is incorrect. The ignition switch is part of the starting system. This component sends a start signal to the starter when the driver activates it.

Answer D is correct. The voltage regulator is part of the charging system. This component controls the output of the charging system as electrical demand changes.

30. Service Consultant A suggests that offering a customer a ride home or to work represents alternative transportation. Service Consultant B suggests that driving the customer to the bus stop is providing alternative transportation. Who is correct?

TASK A.1.5

A. A only
B. B only
C. Both A and B
D. Neither A nor B

Answer A is incorrect. Service Consultant B is also correct.

Answer B is incorrect. Service Consultant A is also correct.

Answer C is correct. Both Service Consultants are correct. Alternative transportation is any way the customer can get where they need to go. It can be the deal breaker for some customers who depend on their vehicle for transportation.

Answer D is incorrect. Both Service Consultants are correct.

31. Service Consultant A says that the radiator is a component of the engine cooling system. Service Consultant B says that the catalytic converter is a component of the ignition system. Who is correct?

TASK B.1.3

A. A only
B. B only
C. Both A and B
D. Neither A nor B

Answer A is correct. Only Service Consultant A is correct. The radiator is part of the engine cooling system. This component is a heat exchanger that releases heat from the hot engine coolant to the outside air.

Answer B is incorrect. The catalytic converter is part of the emissions system. This component is located in the exhaust system. It causes some of the chemicals in the exhaust system to be converted into less toxic substances.

Answer C is incorrect. Only Service Consultant A is correct.

Answer D is incorrect. Service Consultant A is correct.

32. Service Consultant A says that rear-wheel drive vehicles use a half shaft to connect the transmission to the rear axle. Service Consultant B says that front-wheel drive vehicles use a driveshaft with universal joints to connect the transaxle to the drive wheels. Who is correct?

TASK B.2.1, B.2.2

A. A only
B. B only
C. Both A and B
D. Neither A nor B

Answer A is incorrect. Rear-wheel drive vehicles typically use a driveshaft with universal joints to connect the transmission to the rear axle.

Answer B is incorrect. Front-wheel drive vehicles typically use a small shaft (sometimes called a *half shaft*) with constant velocity (CV) joints to connect the transaxle to the drive wheels.

Answer C is incorrect. Neither Service Consultant is correct.

Answer D is correct. Neither Service Consultant is correct. Rear-wheel drive vehicles typically use a driveshaft to transfer power to the rear axle and front-wheel drive vehicles typically use a small shaft (sometimes called a half shaft) to connect the transaxle to the drive wheels.

TASK B.2.2

33. Service Consultant A says that the engine flywheel drives the clutch disc when the clutch pedal is released. Service Consultant B says that the pressure plate is the part of the system that compresses the clutch disc into the flywheel when the clutch pedal is depressed. Who is correct?

A. A only
B. B only
C. Both A and B
D. Neither A nor B

Answer A is correct. Only Service Consultant A is correct. The engine flywheel transfers torque to the clutch friction disc when the clutch pedal is released. The pressure plate holds pressure on this connection in order to allow the transfer of power to occur.

Answer B is incorrect. The pressure plate applies pressure to the clutch disc when the clutch pedal is released.

Answer C is incorrect. Only Service Consultant A is correct.

Answer D is incorrect. Service Consultant A is correct.

TASK A.1.12

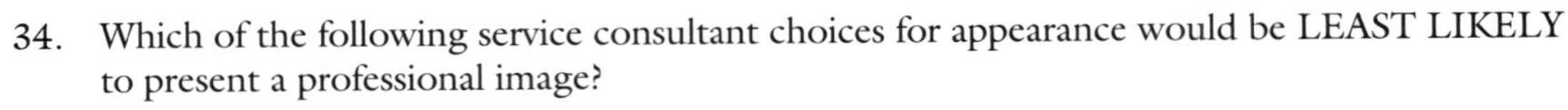

34. Which of the following service consultant choices for appearance would be LEAST LIKELY to present a professional image?

A. Green shirt and blue pants
B. Blue jeans and tennis shoes
C. White shirt and tan pants
D. Clean and neatly groomed hair

Answer A is incorrect. This clothing combination would be an acceptable choice.

Answer B is correct. Wearing blue jeans and tennis shoes is not a good choice for a service consultant to wear. Presenting a professional image is very important in building trust with customers.

Answer C is incorrect. This clothing combination would be an acceptable choice.

Answer D is incorrect. This clothing combination would be an acceptable choice.

TASK B.3.1

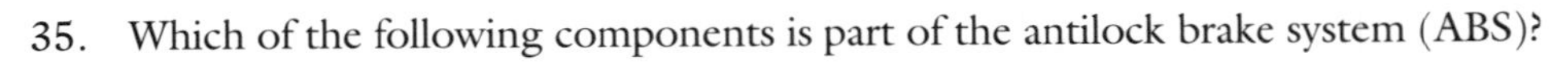

35. Which of the following components is part of the antilock brake system (ABS)?

A. Master cylinder
B. Wheel cylinder
C. Caliper
D. Hydraulic control unit

Answer A is incorrect. The master cylinder is a component of the base hydraulic brake system. This component transfers the linear movement of the brake pedal into hydraulic pressure. This pressurized hydraulic fluid is directed to each wheel to activate either a wheel cylinder or a brake caliper.

Answer B is incorrect. The wheel cylinder is a component of the base hydraulic brake system that uses drum style brakes. This component receives hydraulic pressure from the master cylinder and pushes the brake shoes into the drum to cause the vehicle to slow down.

Answer C is incorrect. The caliper is a component of the base hydraulic brake system that uses disc style brakes. This component receives hydraulic pressure from the master cylinder and compresses the brake pads into the rotor, which causes the vehicle to slow down.

Answer D is correct. The hydraulic control unit is a component of the antilock brake system. This component modulates the brake fluid under heavy braking conditions to prevent wheel lockup.

36. Which component of the steering system connects the steering gear linkage to the steering knuckle?

TASK B.3.2

A. Spring
B. Tie rod end
C. Strut
D. Control arm

Answer A is incorrect. The spring helps support the weight of the body and also compresses when the vehicle is driven over rough terrain.

Answer B is correct. The tie rod end connects the steering gear linkage to the steering knuckle.

Answer C is incorrect. The strut helps to dampen spring activity as the vehicle is driven over varied surfaces.

Answer D is incorrect. The control arm connects the steering knuckle to the vehicle frame or body.

37. Service Consultant A does not recommend that the alignment be checked prior to having new tires installed. Service Consultant B quotes two levels of tires in order to let the customer decide the expense. Who is correct?

TASK B.3.3

A. A only
B. B only
C. Both A and B
D. Neither A nor B

Answer A is incorrect. It is a good idea to advise customers to have the alignment checked prior to purchasing new tires. This practice will result in longer lasting tires for the customers.

Answer B is correct. Only Service Consultant B is correct. It is a good idea to have two quality/price levels of tires quoted in order to let the customer decide how much they can spend.

Answer C is incorrect. Only Service Consultant B is correct.

Answer D is incorrect. Service Consultant B is correct.

38. When a customer objects to the cost of a given repair, the best response by the service consultant would be to:

TASK A.2.6

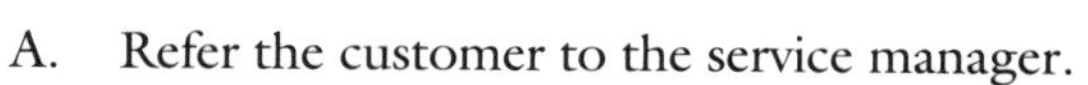

A. Refer the customer to the service manager.
B. Offer the customer a discount to encourage his/her approval.
C. Explain the benefits of having the repair performed.
D. Reschedule the repair for a different time.

Answer A is incorrect. The service consultant should be able to handle most negotiations with customers without involving the service manager.

Answer B is incorrect. The service consultant should not volunteer to discount a repair fee easily. Using logical explanations to explain why the repair is needed is the first choice in negotiations.

Answer C is correct. A service consultant will be most effective in assisting customers who object to repairs by logically explaining why the repair is needed.

Answer D is incorrect. Rescheduling the repair for a different time would not be the first choice in negotiating with a customer. It may sometimes be necessary, however, to have the customer bring the vehicle back at a time when the customer can afford the repair.

TASK B.4.2

39. Service Consultant A says that the evaporator core is the source of heat in the HVAC system. Service Consultant B says that the blend door is a device that controls the temperature of the air that is discharged from the HVAC system? Who is correct?

    A. A only
    B. B only
    C. Both A and B
    D. Neither A nor B

Answer A is incorrect. The source of heat in the HVAC system is the heater core. The evaporator core is the device that cools the air in the HVAC system.

Answer B is correct. Only Service Consultant B is correct. The blend door is located between the evaporator core and the heater core. This device moves when the temperature control lever is adjusted. This component controls duct output temperature by routing the air either through or around the heater core.

Answer C is incorrect. Only Service Consultant B is correct.

Answer D is incorrect. Service Consultant B is correct.

TASK A.2.1

40. Service Consultant A adds the labor total when calculating an estimate for a repair. Service Consultant B adds the parts total when calculating an estimate for a repair. Who is correct?

    A. A only
    B. B only
    C. Both A and B
    D. Neither A nor B

Answer A is incorrect. Service Consultant B is also correct.

Answer B is incorrect. Service Consultant A is also correct.

Answer C is correct. Both Service Consultants are correct. The Service Consultant should add the parts total and the labor total when calculating an estimate. In addition, repair estimates will include sublet repairs charges, chemical charges, disposal charges, and the necessary taxes.

Answer D is incorrect. Both Service Consultants are correct.

TASK B.5.1

41. All of the following procedures are performed during a typical 90,000-mile service EXCEPT:

    A. Replace the spark plugs.
    B. Replace the fuel filter.
    C. Inspect and adjust the brake system.
    D. Replace the A/C compressor.

Answer A is incorrect. The spark plugs are usually replaced during a 90,000-mile service.

Answer B is incorrect. The fuel filter is usually replaced during a 90,000-mile service.

Answer C is incorrect. The brake system is usually inspected and adjusted during a 90,000-mile service.

Answer D is correct. The A/C compressor is not typically replaced during a 90,000-mile service.

42. What is the LEAST LIKELY step that would be followed by a service consultant to verify the accuracy of a repair order prior to calling the customer for repair authorization?

TASK A.1.10

A. Add the labor and the parts up in your mind.

B. Input all the service repair-related data into the computer and verify the availability of the parts.

C. Add the cost of the parts, the cost of the labor, the tax, and the cost of sublet repairs.

D. Input the parts cost, labor cost, tax cost, and the sublet repair cost into a calculator.

Answer A is correct. Adding the labor and parts totals up in your head is not a reliable way to compute accurate estimates for the customer.

Answer B is incorrect. Many computer programs perform automatic calculations to provide accurate repair estimates.

Answer C is incorrect. Adding up all of the elements of the repair process will provide accurate estimates.

Answer D is incorrect. Adding up all of the elements of the repair process on a calculator will provide accurate estimates.

43. Service Consultant A says that maintenance schedules are printed in the vehicle owner's manual. Service Consultant B says that maintenance schedules are selected based on the customer's use of the vehicle. Who is correct?

TASK B.5.3

A. A only

B. B only

C. Both A and B

D. Neither A nor B

Answer A is incorrect. Service Consultant B is also correct.

Answer B is incorrect. Service Consultant A is also correct.

Answer C is correct. Both Service Consultants are correct. The main source of maintenance schedules for the vehicle owner is the owner's manual. Maintenance schedule choice is based on environmental and use parameters. Vehicles that are used under harsher conditions will need to be serviced more frequently than light-duty use.

Answer D is incorrect. Both Service Consultants are correct.

44. All of the following items are needed when collecting vehicle information for the vehicle repair order EXCEPT:

TASK A.1.2

A. Mileage

B. Interior color

C. Vehicle model

D. Vehicle make

Answer A is incorrect. The mileage of the vehicle is a very important piece of information to be collected for the repair order.

Answer B is correct. The interior color of the vehicle is not a common piece of information that would appear on the repair order.

Answer C is incorrect. The vehicle model is a common piece of information that would appear on the repair order.

Answer D is incorrect. The vehicle make is a common piece of information that would appear on the repair order.

TASK B.6.4

45. What is the most likely repair that would be covered by a service contract?

A. Brake pads
B. Wiper blade
C. Water pump
D. Air filter

Answer A is incorrect. Brake pads are usually considered to be wear-type components that would not be covered by a service contract.

Answer B is incorrect. The wiper blades are usually considered to be wear-type components that would not be covered by a service contract.

Answer C is correct. A water pump is a major engine component, which is usually covered by a service contract. Service contracts will vary widely, so the service consultant should read the details of each one before making promises/quotes to the customer.

Answer D is incorrect. The air filter is not usually covered by a service contract because it is considered a maintenance item.

TASK B.7.1

46. Which VIN digit represents the vehicle year?

A. First
B. Sixth
C. Eighth
D. Tenth

Answer A is incorrect. The first VIN digit is the country of origin.

Answer B is incorrect. The sixth VIN digit is the body code.

Answer C is incorrect. The eighth VIN digit is the engine code.

Answer D is correct. The tenth digit is the vehicle year.

TASK B.7.3

47. Service Consultant A uses the internet to locate pictures of the major vehicle systems in order to show the customer which items need service. Service Consultant B keeps a notebook with various pictures of vehicle systems at the service desk to show the customer which items need service. Who is correct?

A. A only
B. B only
C. Both A and B
D. Neither A nor B

Answer A is incorrect. Service Consultant B is also correct.

Answer B is incorrect. Service Consultant A is also correct.

Answer C is correct. Both Service Consultants are correct. Using the internet or a printed picture can assist the service consultant in showing the customer which items need service.

Answer D is incorrect. Both Service Consultants are correct.

48. Service Consultant A says that it is very helpful to know what has been done to the vehicle in the past. Service Consultant B says that some shops have computerized software that stores the service data on each vehicle that has been in the repair facility. Who is correct?

TASK A.1.7

A. A only

B. B only

C. Both A and B

D. Neither A nor B

Answer A is incorrect. Service Consultant B is also correct.

Answer B is incorrect. Service Consultant A is also correct.

Answer C is correct. Both Service Consultants are correct. Having the repair history is extremely helpful when planning the service and maintenance on a customer's vehicle. Many shops use computerized software that stores the service data on each vehicle that was serviced by the shop. This type of data storage is very valuable to use when searching for prior repairs that have been performed.

Answer D is incorrect. Both Service Consultants are correct.

49. Service Consultant A says that some shops sublet body repairs to a body shop. Service Consultant B says that sublet repairs should be added to the repair order when adding up the total bill for the customer. Who is correct?

TASK C.2

A. A only

B. B only

C. Both A and B

D. Neither A nor B

Answer A is incorrect. Service Consultant B is also correct.

Answer B is incorrect. Service Consultant A is also correct.

Answer C is correct. Both Service Consultants are correct. Many auto repair shops do not have a body shop in-house, so body-related repairs must be sent to a body shop. All sublet repairs should be added to the repair order when adding the total of the repair bill for the customer.

Answer D is incorrect. Both Service Consultants are correct.

50. An upset customer comes in when the service department is very busy with a complaint about past repair service. Service Consultant A listens to the customer until he has the opportunity to show that he/she is wrong. Service Consultant B lets the customer tell the whole story. Who is correct?

TASK C.4

A. A only

B. B only

C. Both A and B

D. Neither A nor B

Answer A is incorrect. The service consultant should not be confrontational with an upset customer. The service consultant should just listen until the customer has had a chance to present his or her complaints.

Answer B is correct. Only Service Consultant B is correct. It is a good practice to allow the customer to complete his or her story about the problem with the vehicle.

Answer C is incorrect. Only Service Consultant B is correct.

Answer D is incorrect. Service Consultant B is correct.

PREPARATION EXAM 6—ANSWER KEY

1. A
2. A
3. B
4. C
5. C
6. A
7. A
8. B
9. C
10. B
11. C
12. C
13. B
14. B
15. C
16. C
17. C
18. D
19. A
20. C
21. C
22. B
23. A
24. B
25. C
26. B
27. A
28. B
29. B
30. C
31. C
32. C
33. B
34. C
35. B
36. D
37. B
38. A
39. A
40. B
41. C
42. C
43. A
44. D
45. A
46. C
47. A
48. A
49. D
50. B

PREPARATION EXAM 6—EXPLANATIONS

TASK A.1.1

1. A potential customer calls very concerned about an estimate received from another shop. Which of the following should the service consultant do?

   A. Show concern for the potential customer and offer an appointment for a second opinion.

   B. Research the amount that service consultant's shop would charge for that service.

   C. Offer a discount if the potential customer brings the vehicle in.

   D. Tell the potential customer that the other shop is dishonest.

   Answer A is correct. A service consultant should show concern for the potential customer and then offer an appointment for a second opinion. This will allow the repair shop to potentially gain a customer if good service is provided.

   Answer B is incorrect. It would not be wise to quote a price for a repair prior to fully diagnosing the problem and then assembling an accurate estimate.

   Answer C is incorrect. A service consultant should not volunteer to discount the repair without reason or complete diagnosis.

   Answer D is incorrect. A service consultant should not express a negative opinion about other shops in front of the customer.

2. Which fuel system component is mounted in the fuel tank?

TASK B.1.1

   A. Fuel pump
   B. Fuel supply line
   C. Fuel injector
   D. Fuel level gauge

   Answer A is correct. The fuel pump is located in the fuel tank on fuel-injected vehicles.

   Answer B is incorrect. The fuel supply line connects the fuel tank to the fuel rail.

   Answer C is incorrect. The fuel injectors are located on the intake manifold.

   Answer D is incorrect. The fuel level gauge is located in the instrument cluster inside the cabin area.

3. Which of the following items would be typically covered under a manufacturer's power train warranty?

TASK B.6.1

   A. Tires
   B. Transmission
   C. Brake pads
   D. Fuel filter

   Answer A is incorrect. The tires are considered wear items and would not be covered under a power train warranty.

   Answer B is correct. The transmission would be considered a power train component and, therefore, would likely be covered under this type of warranty.

   Answer C is incorrect. The brake pads are considered wear items and would not be covered under a power train warranty.

   Answer D is incorrect. The fuel filter would not typically be covered under a power train warranty.

4. Service Consultant A says that customers expect to be treated with respect when they arrive at the repair shop. Service Consultant B says that customers expect a sincere service-oriented attitude when they are doing business with a repair shop. Who is correct?

TASK A.1.4

   A. A only
   B. B only
   C. Both A and B
   D. Neither A nor B

   Answer A is incorrect. Service Consultant B is also correct.

   Answer B is incorrect. Service Consultant A is also correct.

   Answer C is correct. Both Service Consultants are correct. The service consultant should always have a good attitude and treat his customers with respect when interacting with them at the repair shop.

   Answer D is incorrect. Both Service Consultants are correct.

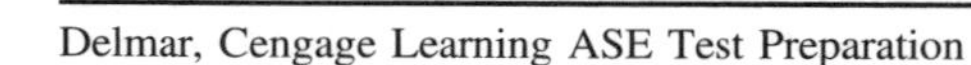

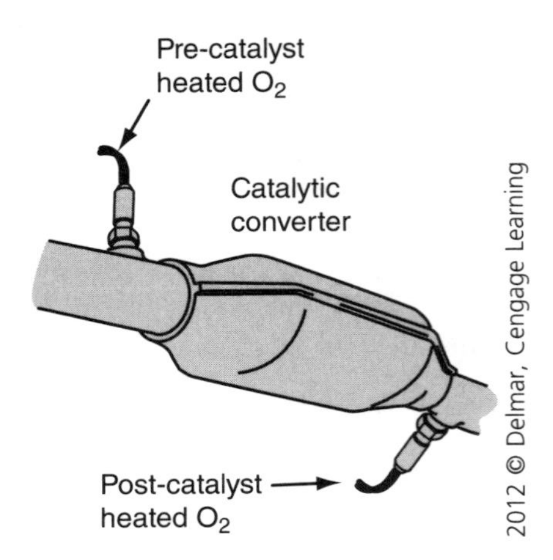

TASK B.1.2

5. Referring to the figure above, Service Consultant A says that this vehicle uses an OBD-II emission system. Service Consultant B says that the post-catalyst O_2 sensor measures the efficiency of the catalytic converter. Who is correct?

    A. A only
    B. B only
    C. Both A and B
    D. Neither A nor B

Answer A is incorrect. Service Consultant B is also correct.

Answer B is incorrect. Service Consultant A is also correct.

Answer C is correct. Both Service Consultants are correct. All OBD-II vehicles will have O_2 sensors before and after the catalytic converter. The rear O_2 sensor provides efficiency data to the engine computer about the catalytic converter.

Answer D is incorrect. Both Service Consultants are correct.

TASK A.1.6

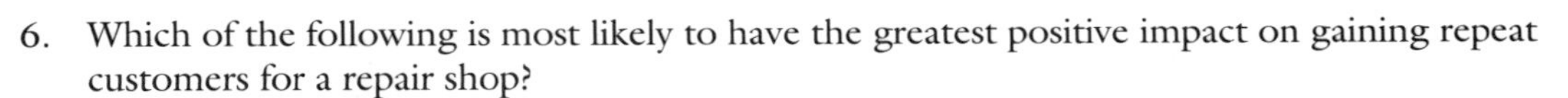

6. Which of the following is most likely to have the greatest positive impact on gaining repeat customers for a repair shop?

    A. Receiving an honest and fair service at a reasonable price
    B. The service consultant's appearance
    C. Printed advertisement in a local newspaper
    D. Discount pricing

Answer A is correct. Providing an honest and fair service at a reasonable price will have a great positive impact on gaining repeat customers for a repair shop.

Answer B is incorrect. The service consultant should strive to have a professional appearance, but this practice will not have a large impact on gaining repeat customers for a repair shop.

Answer C is incorrect. Using various marketing tools like advertising in a local newspaper will help increase business, but it is not the greatest factor in gaining repeat customers for a repair shop.

Answer D is incorrect. Using discount methods will assist in increasing the customer count for a shop, but it is not the greatest factor in gaining repeat customers.

7. A customer walks into the repair shop at 4:30 p.m. requesting an oil change and a tire rotation. The estimated time for this repair is 45 minutes and the shop closes at 5:30 p.m. A service technician is available to perform this service. Service Consultant A quickly completes the repair order and promises that the car will be completed that same day before the close of business. Service Consultant B recommends that the customer reschedule this repair due to the limited time left in the day. Who is correct?

TASK C.1

A. A only

B. B only

C. Both A and B

D. Neither A nor B

Answer A is correct. Only Service Consultant A is correct. The service consultant should quickly get this vehicle in and processed because it can easily be completed before the end of the day.

Answer B is incorrect. There is not a good reason given in the scenario for turning the customer away.

Answer C is incorrect. Only Service Consultant A is correct.

Answer D is incorrect. Service Consultant A is correct.

8. A vehicle in the shop for an oil change shows approximately 59,000 miles on the odometer. What should the service consultant do?

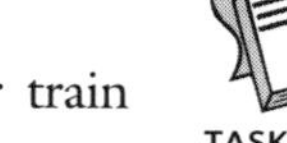

TASK A.1.8

A. Advise the customer that the 60,000-mile service is covered under the power train warranty.

B. Quote a price and offer to schedule an appointment for a 60,000-mile service.

C. Offer a discount if the customer schedules the appointment within the next week.

D. Advise the customer that the 60,000-mile service is critical and the vehicle is not safe to drive if the service is not performed.

Answer A is incorrect. Maintenance services are not typically covered under a power train warranty.

Answer B is correct. The service consultant should be ready to quote a price for upcoming services for each customer. Using this practice will help the customer plan for the upcoming expense, as well as to plan a time for dropping off his/her vehicle.

Answer C is incorrect. The service consultant should not offer discounts randomly.

Answer D is incorrect. The service consultant should not use scare tactics to influence the customer's decisions about his/her car maintenance.

9. Which of the following conditions will cause the tire pressure warning light to illuminate?

TASK B.3.1, B.5.2

A. Tire pressure that is three to six pounds above the specification

B. Tires that are worn below 4/32-inch tread depth

C. Tire pressure that drops four to six pounds below the specification

D. Tires that are out of balance

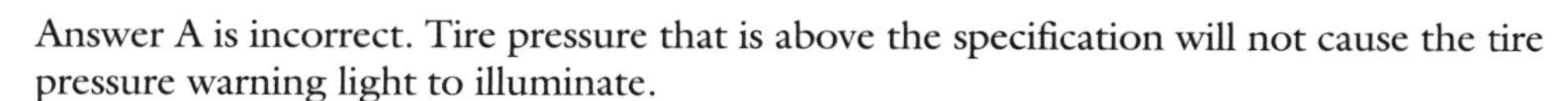

Answer A is incorrect. Tire pressure that is above the specification will not cause the tire pressure warning light to illuminate.

Answer B is incorrect. Worn tires will not cause the tire pressure light to illuminate.

Answer C is correct. Tire pressure that drops four to six pounds below the specification will cause the tire pressure warning light to illuminate.

Answer D is incorrect. Out-of-balance tires will not cause the tire pressure warning light to illuminate.

TASK A.1.12

10. Service Consultant A wears his/her uniform without the shirt tucked in. Service Consultant B wears a casual uniform to work. Who is correct?

    A. A only
    B. B only
    C. Both A and B
    D. Neither A nor B

Answer A is incorrect. The service consultant should tuck in his/her shirt while at work. Appearance is an important aspect in promoting a good customer experience at a repair shop.

Answer B is correct. Only Service Consultant B is correct. Wearing some type of casual uniform presents a professional appearance that will create a good impression for the customer.

Answer C is incorrect. Only Service Consultant B is correct.

Answer D is incorrect. Service Consultant B is correct.

TASK B.1.1

11. Which engine component rotates and causes the valve to open at the correct time?

    A. Crankshaft
    B. Timing gear
    C. Camshaft
    D. Flywheel

Answer A is incorrect. The crankshaft is connected to the pistons, which create the rotational force during combustion.

Answer B is incorrect. The timing gear is driven by a belt or chain to connect the crankshaft and camshaft.

Answer C is correct. The camshaft rotates and has eccentrics that cause the valve to open at the correct time. Many late-model engines have overhead cam engine designs.

Answer D is incorrect. The flywheel is on the back of the engine and is driven by the crankshaft.

TASK A.2.1

12. Service Consultant A adds the miscellaneous expense total when calculating an estimate for a repair. Service Consultant B adds the sublet repair total when calculating an estimate for a repair. Who is correct?

    A. A only
    B. B only
    C. Both A and B
    D. Neither A nor B

Answer A is incorrect. Service Consultant B is also correct.

Answer B is incorrect. Service Consultant A is also correct.

Answer C is correct. Both Service Consultants are correct. All of the varied elements of the repair order should be calculated when putting an estimate together for a customer.

Answer D is incorrect. Both Service Consultants are correct.

13. Service Consultant A says that the starter relay is part of the charging system. Service Consultant B says that the voltage regulator is part of the charging system. Who is correct?

TASK B.1.1, B.4.3

A. A only
B. B only
C. Both A and B
D. Neither A nor B

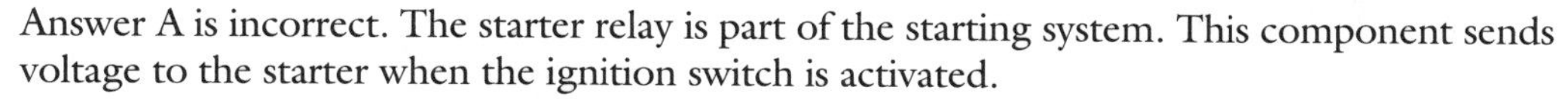

Answer A is incorrect. The starter relay is part of the starting system. This component sends voltage to the starter when the ignition switch is activated.

Answer B is correct. Only Service Consultant B is correct. The voltage regulator is part of the charging system. This component controls the output of the alternator as the electrical load varies.

Answer C is incorrect. Only Service Consultant B is correct.

Answer D is incorrect. Service Consultant B is correct.

14. Service Consultant A refers the customer to the service manager if the customer asks detailed questions about why a particular service is being recommended. Service Consultant B provides clear and understandable answers to detailed questions that his customers have. Who is correct?

TASK A.2.3

A. A only
B. B only
C. Both A and B
D. Neither A nor B

Answer A is incorrect. The service consultant should be able to answer detailed questions about a recommended service.

Answer B is correct. Only Service Consultant B is correct. The service consultant should always provide clear and understandable answers to all questions that customers have about their vehicle.

Answer C is incorrect. Only Service Consultant B is correct.

Answer D is incorrect. Service Consultant B is correct.

15. Service Consultant A monitors the weekly flat-rate hours that each technician flags to determine the efficiency of the technicians. Service Consultant B monitors the total shop hours for each month to determine shop productivity. Who is correct?

TASK A.3.7

A. A only
B. B only
C. Both A and B
D. Neither A nor B

Answer A is incorrect. Service Consultant B is also correct.

Answer B is incorrect. Service Consultant A is also correct.

Answer C is correct. Both Service Consultants are correct. Weekly flat-rate hours is a key indicator for determining technician efficiency and total shop hours flagged is a key indicator for determining shop productivity.

Answer D is incorrect. Both Service Consultants are correct.

TASK A.2.5

16. Service Consultant A says that the benefit of having regular oil changes is reduced engine wear. Service Consultant B says the benefit of replacing the air filter at regular intervals is reduced engine wear. Who is correct?

 A. A only
 B. B only
 C. Both A and B
 D. Neither A nor B

 Answer A is incorrect. Service Consultant B is also correct.

 Answer B is incorrect. Service Consultant A is also correct.

 Answer C is correct. Both Service Consultants are correct. Changing the oil and replacing the air filter will help reduce the engine wear on all vehicles.

 Answer D is incorrect. Both Service Consultants are correct.

TASK B.4.1

17. Service Consultant A says that the negative battery cable connects to the engine block. Service Consultant B says that the positive battery cable connects to the starter solenoid. Who is correct?

 A. A only
 B. B only
 C. Both A and B
 D. Neither A nor B

 Answer A is incorrect. Service Consultant B is also correct.

 Answer B is incorrect. Service Consultant A is also correct.

 Answer C is correct. Both Service Consultants are correct. The negative battery cable does connect to the engine block and the positive battery cable does connect to the "bat" connection at the starter solenoid.

 Answer D is incorrect. Both Service Consultants are correct.

TASK A.2.2

18. Which of the following repairs would be the highest priority for a customer to repair in relation to the safety of the vehicle?

 A. Replacing the power window motor
 B. Replacing the blower motor
 C. Replacing the radio
 D. Replacing the clock spring for the airbag inflator

 Answer A is incorrect. Replacing the power window motor is not a safety-related repair.

 Answer B is incorrect. Replacing the blower motor is not a safety-related repair.

 Answer C is incorrect. Replacing the radio is not a safety-related repair.

 Answer D is correct. The clock spring for the airbag inflator is a safety-related component and is a high-priority repair.

19. A customer brings a car to the repair shop with a complaint that the steering wheel vibrates while braking. Service Consultant A recommends having the brake technician diagnose the problem. Service Consultant B recommends that the rotors will need to be turned or replaced. Who is correct?

TASK A.2.3, A.3.2

A. A only
B. B only
C. Both A and B
D. Neither A nor B

Answer A is correct. Only Service Consultant A is correct. The service consultant should have the brake technician diagnose this concern and then offer a quote to complete the repair.

Answer B is incorrect. The service consultant should refrain from giving an opinion about the possible problems that a vehicle might have.

Answer C is incorrect. Only Service Consultant A is correct.

Answer D is incorrect. Service Consultant A is correct.

20. A customer arrives at the repair shop requesting an oil change and a tire rotation. Service Consultant A recommends that the shop perform a brake cleaning and inspection while the wheels are removed. Service Consultant B recommends that the air filter be inspected while the vehicle is at the shop. Who is correct?

TASK A.2.4

A. A only
B. B only
C. Both A and B
D. Neither A nor B

Answer A is incorrect. Service Consultant B is also correct.

Answer B is incorrect. Service Consultant A is also correct.

Answer C is correct. Both Service Consultants are correct. When a vehicle is in for maintenance items, it is advisable to use this opportunity to inspect related items to increase the level of service for the customer.

Answer D is incorrect. Both Service Consultants are correct.

21. Service Consultant A says that providing an accurate estimate is a useful tool for closing a sale. Service Consultant B says that explaining the warranty on the repair parts is a good way to close a sale. Who is correct?

TASK A.2.6

A. A only
B. B only
C. Both A and B
D. Neither A nor B

Answer A is incorrect. Service Consultant B is also correct.

Answer B is incorrect. Service Consultant A is also correct.

Answer C is correct. Both Service Consultants are correct. Service consultants should always provide accurate estimates in addition to explaining the warranty on the parts that will be installed during the service.

Answer D is incorrect. Both Service Consultants are correct.

TASK A.3.2

22. A technician turns in a repair order that recommends replacement of the fuel filter. Service Consultant A calls the customer to get approval to complete the repair. Service Consultant B consults with the technician to find out the reason for filter replacement. Who is correct?

    A. A only
    B. B only
    C. Both A and B
    D. Neither A nor B

Answer A is incorrect. The service consultant should investigate the reason that the fuel filter needs to be replaced because the customer will likely inquire about it.

Answer B is correct. Only Service Consultant B is correct. Having a full understanding about why a component needs changing will assist the service consultant in making the sale of the item and related service.

Answer C is incorrect. Only Service Consultant B is correct.

Answer D is incorrect. Service Consultant B is correct.

TASK A.3.5

23. Which of the following is the most likely method to be used when checking for quality control in a repair shop?

    A. Keep a record of repeat repairs for each technician.
    B. Monitor the total shop revenue each month.
    C. Keep a record of redeemed coupons for each month.
    D. Keep a record of random maintenance customers for each month.

Answer A is correct. Keeping a record of repeat repairs for each technician is a way to monitor quality control for the shop.

Answer B is incorrect. The total revenue numbers for the shop would not provide any indication of the quality of the repairs being sold.

Answer C is incorrect. The total of the redeemed coupons would not provide any indication of the quality of the repairs being sold.

Answer D is incorrect. Having a record of the maintenance being performed will not provide any indication of the quality of the repairs being sold.

TASK A.2.4

24. All of the following are benefits of recommending overdue services to a current customer who has left his/her car at your shop EXCEPT:

    A. The customer will begin to trust that the shop is looking out for his/her well-being and safety.
    B. The employees will have to work late to complete the repair.
    C. The shop is more profitable from the increased sale of labor.
    D. The shop is more profitable from the increased sale of parts.

Answer A is incorrect. Gaining the trust of the customer is a benefit that results from recommending overdue services on a vehicle.

Answer B is correct. Causing the employees to work late is not a benefit that results from recommending overdue services on a vehicle.

Answer C is incorrect. Increasing the labor profit for the shop is a benefit that results from recommending overdue services on a vehicle.

Answer D is incorrect. Increasing the parts profit for the shop is a benefit that results from recommending overdue services on a vehicle.

25. Service Consultant A meets with the service manager each week to discuss the stresses involved with the job. Service Consultant B continuously monitors the shop environment by discussing the daily activities with the shop technicians. Who is correct?

TASK A.3.8

    A. A only
    B. B only
    C. Both A and B
    D. Neither A nor B

Answer A is incorrect. Service Consultant B is also correct.

Answer B is incorrect. Service Consultant A is also correct.

Answer C is correct. Both Service Consultants are correct. Communication is an important activity for a good service consultant to remain effective. She should be in continuous communication with the service manager and the technicians to do a quality job and promote organizational effectiveness.

Answer D is incorrect. Both Service Consultants are correct.

26. When writing up a customer's work order, which of the following is the first thing to ask for?

TASK A.1.2

    A. The vehicle identification number (VIN)
    B. The customer's name
    C. The main customer concern
    D. The license plate number

Answer A is incorrect. Most customers will not know what their VIN is nor where it is located, let alone be able to dictate it to you.

Answer B is correct. Trainers and customer service people all agree that the first thing to do is find out the name of the person you are talking to. The person will subsequently feel more comfortable throughout the transaction.

Answer C is incorrect. Although the customer's main concern is the reason the customer has brought the vehicle in, it actually ranks second or third in your list of items to collect.

Answer D is incorrect. Most people do not know their license number.

27. Service Consultant A says that the crankshaft sensor sends engine speed data to the electronic control module (ECM). Service Consultant B says that the crankshaft sensor needs to be mounted near the drive belt. Who is correct?

TASK B.1.1

    A. A only
    B. B only
    C. Both A and B
    D. Neither A nor B

Answer A is correct. Only Service Consultant A is correct. The crankshaft sensor sends engine speed data to the ECM. This component picks up a signal from a reluctor ring located on the crankshaft.

Answer B is incorrect. The crankshaft sensor has to be mounted near the crankshaft reluctor ring.

Answer C is incorrect. Only Service Consultant A is correct.

Answer D is incorrect. Service Consultant A is correct.

TASK B.1.2

28. Service Consultant A says that the alternator maintains belt tension on a serpentine belt system. Service Consultant B says that a serpentine belt is wider than a V-belt. Who is correct?

A. A only

B. B only

C. Both A and B

D. Neither A nor B

Answer A is incorrect. The belt tensioner provides the tension on most vehicles when serpentine belts are used.

Answer B is correct. Only Service Consultant B is correct. Serpentine belts are wide and flat and are typically held tight by a device called a *tensioner*.

Answer C is incorrect. Only Service Consultant B is correct.

Answer D is incorrect. Service Consultant B is correct.

TASK B.1.3

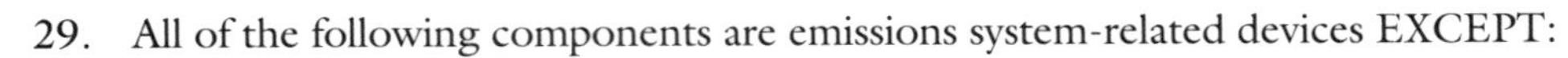

29. All of the following components are emissions system-related devices EXCEPT:

A. Knock sensor

B. Cabin temperature sensor

C. Crankshaft sensor

D. Oxygen sensor

Answer A is incorrect. The knock sensor is part of the emissions system. This component is mounted on the engine and creates a voltage signal when detonation is present in the engine.

Answer B is correct. The cabin temperature sensor is not part of the emissions system. This component is part of the HVAC system that provides cabin temperature data for the HVAC computer.

Answer C is incorrect. The crankshaft sensor is part of the emissions system. This component is mounted near the engine crankshaft and sends a speed signal to the engine computer.

Answer D is incorrect. The oxygen sensor is part of the emissions system. This component is mounted in the exhaust system and sends signals as the oxygen levels vary in the exhaust stream.

TASK A.1.5

30. Service Consultant A says that the some customers will ask for a ride back to their house when they make their appointment. Service Consultant B says that some customers will ask for a ride back to their work location when they make their appointment. Who is correct?

A. A only

B. B only

C. Both A and B

D. Neither A nor B

Answer A is incorrect. Service Consultant B is also correct.

Answer B is incorrect. Service Consultant A is also correct.

Answer C is correct. Both Service Consultants are correct. The service consultant should be prepared for the customer to request a ride back to his/her preferred location.

Answer D is incorrect. Both Service Consultants are correct.

31. Service Consultant A says that rear-wheel drive vehicles use a driveshaft with universal joints on each end to connect the transmission to the rear axle. Service Consultant B says that front-wheel drive vehicles use a half shaft with constant velocity (CV) joints to connect the transaxle to the drive wheels. Who is correct?

TASK B.2.1, B.2.2

A. A only
B. B only
C. Both A and B
D. Neither A nor B

Answer A is incorrect. Service Consultant B is also correct.

Answer B is incorrect. Service Consultant A is also correct.

Answer C is correct. Both Service Consultants are correct. Rear-wheel drive vehicles use a driveshaft and front-wheel drive vehicles use a half shaft.

Answer D is incorrect. Both Service Consultants are correct.

32. Service Consultant A says that the clutch disc is splined onto the transmission input shaft. Service Consultant B says that the pressure plate is the part of the system that compresses the clutch disc into the flywheel when the clutch pedal is released. Who is correct?

TASK B.2.2

A. A only
B. B only
C. Both A and B
D. Neither A nor B

Answer A is incorrect. Service Consultant B is also correct.

Answer B is incorrect. Service Consultant A is also correct.

Answer C is correct. Both Service Consultants are correct. The clutch disc is splined onto the transmission input shaft and transfers engine rotation into the transmission when the clutch pedal is released. The pressure plate provides the force that holds pressure on the clutch disc.

Answer D is incorrect. Both Service Consultants are correct.

33. Service Consultant A says that the automatic transmission fluid should be changed more frequently if the vehicle is driven at highway speeds most of the time. Service Consultant B says that the transmission fluid is cooled by routing it into the transmission oil cooler, which is mounted inside the radiator. Who is correct?

TASK B.2.3

A. A only
B. B only
C. Both A and B
D. Neither A nor B

Answer A is incorrect. The automatic transmission fluid will need to be changed more frequently on vehicles that pull a heavy load on a regular basis due to the increased operating temperature of the fluid.

Answer B is correct. Only Service Consultant B is correct. The transmission fluid gets cooled by directing it to the transmission oil cooler. This oil cooler is mounted inside the radiator on most vehicles. Vehicles set up with a towing package will usually have another oil cooler that is mounted in front of the radiator.

Answer C is incorrect. Only Service Consultant B is correct.

Answer D is incorrect. Service Consultant B is correct.

TASK A.1.14

34. A customer arrives to pick up his vehicle after the transmission has been replaced. Service Consultant A carefully explains all of the repairs and charges and asks the customer if there are any questions. Service Consultant B calls the service porter to pull the vehicle around to the pickup area after answering all of the customer's questions. Who is correct?

A. A only
B. B only
C. Both A and B
D. Neither A nor B

Answer A is incorrect. Service Consultant B is also correct.

Answer B is incorrect. Service Consultant A is also correct.

Answer C is correct. Both Service Consultants are correct. The service consultant should make the vehicle pick-up process as convenient and informative as possible for the customer. The customer may have questions about the repair, so the service consultant should take the time needed to answer all of the questions well.

Answer D is incorrect. Both Service Consultants are correct.

TASK B.3.1

35. What is the most likely location for the antilock brake system (ABS) hydraulic modulator?

A. Near the wheel speed sensor
B. Inside the engine compartment
C. Inside the passenger compartment
D. Next to the fuel tank

Answer A is incorrect. The wheel speed sensors are located near each wheel. The ABS hydraulic modulator is mounted in the engine compartment.

Answer B is correct. The ABS hydraulic modulator is typically mounted inside the engine compartment area. This component modulates the brake fluid during panic stops to prevent total wheel lockup.

Answer C is incorrect. The ABS hydraulic modulator is not located in the passenger compartment.

Answer D is incorrect. The ABS hydraulic modulator is not typically located near the fuel tank.

TASK B.3.2

36. Which component of the suspension system connects the steering knuckle to the vehicle body?

A. Spring
B. Tie rod end
C. Strut
D. Control arm

Answer A is incorrect. The spring supports the weight of the body and compresses when uneven surfaces are driven over.

Answer B is incorrect. The tie rod end connects the steering gear linkage to the steering knuckle.

Answer C is incorrect. The strut is a component that reduces the spring compressions of the suspension system.

Answer D is correct. The control arm connects the steering knuckle to the vehicle body.

37. Service Consultant A recommends running the tire pressure 15 psi higher than the specification in order to increase tire life. Service Consultant B recommends rotating the tires every 7,500 miles in order to increase tire life. Who is correct?

TASK B.3.3

A. A only
B. B only
C. Both A and B
D. Neither A nor B

Answer A is incorrect. It is not advisable to run the tire pressure at 15 psi higher than specification. This practice will cause poor ride quality and decrease handling performance.

Answer B is correct. Only Service Consultant B is correct. Rotating the vehicle tires on a regular basis will increase the life of the tires.

Answer C is incorrect. Only Service Consultant B is correct.

Answer D is incorrect. Service Consultant B is correct.

38. A customer is objecting to the cost of the repair bill when picking up her vehicle. The repair had been authorized by her husband prior to completing the repair. Service Consultant A asks the customer to call her husband to verify his authorization. Service Consultant B refers the customer to the service manager to find a solution. Who is correct?

TASK A.2.6

A. A only
B. B only
C. Both A and B
D. Neither A nor B

Answer A is correct. Only Service Consultant A is correct. The service consultant can likely clear up this problem by having the customer call her husband to verify his approval.

Answer B is incorrect. The service consultant should try to handle any objection from a customer prior to involving the service manager.

Answer C is incorrect. Only Service Consultant A is correct.

Answer D is incorrect. Service Consultant A is correct.

39. Service Consultant A says that the heater core is the source of heat in the HVAC system. Service Consultant B says that the blend door is the device that controls the location to which the air is discharged in the HVAC system? Who is correct?

TASK B.4.2

A. A only
B. B only
C. Both A and B
D. Neither A nor B

Answer A is correct. Only Service Consultant A is correct. The heater core is the source of heat for the HVAC system. This component is a heat exchanger that is located in the duct system and has hot engine coolant running through it.

Answer B is incorrect. The blend door controls the temperature of the air being distributed by the duct system. The mode doors control the location from which the air comes out of the HVAC system.

Answer C is incorrect. Only Service Consultant A is correct.

Answer D is incorrect. Service Consultant A is correct.

TASK A.2.2

40. Which of the following repair procedures would be the LEAST LIKELY to be considered a high priority repair?

    A. Replacement of a worn tie rod end
    B. Brake fluid flush and fill
    C. Replacement of a tire with the steel showing
    D. Replacement of worn front brake pads

    Answer A is incorrect. A worn tie rod end is a high priority repair, because the vehicle could lose control if the tie rod fails.

    Answer B is correct. Flushing the brake fluid would not be considered a high priority repair due to the low possibility that it could cause immediate safety problems for the vehicle.

    Answer C is incorrect. A tire with the steel showing is a high priority repair because the tire could blow out at any time and cause the vehicle to crash.

    Answer D is incorrect. Brake pads that are worn should be replaced because they could fail and cause the vehicle to crash.

TASK B.5.1

41. All of the following procedures would be completed during a 90,000-mile service EXCEPT:

    A. Drain and fill the automatic transmission fluid.
    B. Replace the spark plugs.
    C. Replace the alternator.
    D. Replace the fuel filter.

    Answer A is incorrect. The transmission would be drained and filled during a 90,000-mile service.

    Answer B is incorrect. The spark plugs would typically be replaced during a 90,000-mile service.

    Answer C is correct. The alternator is not typically replaced during a 90,000-mile service.

    Answer D is incorrect. The fuel filter is typically replaced during a 90,000-mile service.

TASK A.1.9, B.3.2, B.3.5

42. Service Consultant A shares the expected completion time with the technician so that completion expectations are clear. Service Consultant B asks the technician to alert him if the completion time changes due to a problem in the repair. Who is correct?

    A. A only
    B. B only
    C. Both A and B
    D. Neither A nor B

    Answer A is incorrect. Service Consultant B is also correct.

    Answer B is incorrect. Service Consultant A is also correct.

    Answer C is correct. Both Service Consultants are correct. The service consultant should always keep the technician informed about the expected completion times on each vehicle and the technician should alert the service consultant if a problem is encountered that will change the estimated completion time.

    Answer D is incorrect. Both Service Consultants are correct.

43. Service Consultant A says that maintenance schedules are printed in the vehicle owner's manual. Service Consultant B says that maintenance schedules are the same for all types of driving styles. Who is correct?

TASK B.5.3

A. A only
B. B only
C. Both A and B
D. Neither A nor B

Answer A is correct. Only Service Consultant A is correct. The vehicle owner's manual typically contains a maintenance schedule for each driving style that the vehicle may experience.

Answer B is incorrect. Most manufacturers vary maintenance schedules depending on how the vehicle is driven.

Answer C is incorrect. Only Service Consultant A is correct.

Answer D is incorrect. Service Consultant A is correct.

44. All of the following pieces of customer information might be included on a repair order EXCEPT:

TASK A.1.2

A. Customer's email address
B. Vehicle make and model
C. Cell phone number
D. Service consultant's cell phone number

Answer A is incorrect. The customer's email address is a valuable piece of information for a repair shop to have on the repair order. Email is a popular method of communication with customers.

Answer B is incorrect. The vehicle make and model are important pieces of information to have on the repair order.

Answer C is incorrect. The customer's cell phone number is an important piece of information to have on the repair order. This is likely the best way to communicate with the customers to gain approval for repairs, as well as updating them about their vehicle.

Answer D is correct. The service consultant's phone number is not a likely item to be included on a repair order. However, the service consultant might have her business cell phone number listed on her business card.

45. Service Consultant A says that a manufacturer technical service bulletin (TSB) is a technical document created to help technicians repair pattern failures more quickly. Service Consultant B says that a power train warranty is a program that manufacturers create to invite customers to bring their vehicles back in for a free repair in order to correct a safety fault in the vehicle. Who is correct?

TASK B.6.3

A. A only
B. B only
C. Both A and B
D. Neither A nor B

Answer A is correct. Only Service Consultant A is correct. A TSB is a technical document that manufacturers create to assist technicians in repairing pattern failures in their vehicles.

Answer B is incorrect. A recall is a program that manufacturers create to invite customers back in for a free safety-related repair in the vehicle.

Answer C is incorrect. Only Service Consultant A is correct.

Answer D is incorrect. Service Consultant A is correct.

TASK B.7.1

46. Which VIN digit represents the vehicle engine?

    A. First
    B. Sixth
    C. Eighth
    D. Tenth

Answer A is incorrect. The first VIN digit is the country of origin.

Answer B is incorrect. The sixth VIN digit is the body code.

Answer C is correct. The eighth VIN digit is the engine code.

Answer D is incorrect. The tenth VIN digit is the vehicle year.

TASK B.7.1, B.7.4

47. Service Consultant A says that the vehicle identification number has a digit that reveals the model of the vehicle. Service Consultant B says that a sedan is a car that has only two doors. Who is correct?

    A. A only
    B. B only
    C. Both A and B
    D. Neither A nor B

Answer A is correct. Only Service Consultant A is correct. The vehicle identification number does have a digit that reveals the model of the vehicle. Typically, the fourth or fifth digit designates the model of the vehicle.

Answer B is incorrect. A sedan is a car that has four doors.

Answer C is incorrect. Only Service Consultant A is correct.

Answer D is incorrect. Service Consultant A is correct.

TASK A.1.8, A.2.4

48. What is the most likely reason to suggest additional repair work on a customer's vehicle?

    A. The responsibility of the repair shop to advise the customer of needed service
    B. To increase the amount of labor charges for the shop
    C. To increase the amount of parts sold by the shop
    D. To help customers maintain vehicle value by having needed repair work done

Answer A is correct. Quality service repair shops should strive to identify services that the customers' vehicles need. This builds trust with customers by showing them that the repair shop is carefully inspecting their vehicles for needed services. Care should be taken to only recommend the things that are necessary and not just items that would increase profit.

Answer B is incorrect. Increased labor sales are a result of recommending additional repair work, but the shop should have the customer's safety and trust in mind when making these suggestions.

Answer C is incorrect. Increased parts sales result from additional service work. However, the shop should use honest practices when recommending additional services.

Answer D is incorrect. A well-maintained vehicle retains more value than one that is poorly maintained, but this is not the main reason to suggest these services to the customer.

49. Service Consultant A schedules appointments over the phone and sometimes forgets to add them to the appointment log. Service Consultant B invites customers that he meets after work to come to the repair shop without adding them to the appointment log. Who is correct?

TASK C.1, C.3

A. A only
B. B only
C. Both A and B
D. Neither A nor B

Answer A is incorrect. The service consultant should be diligent in adding all appointments to the appointment log to prevent problematic surprises of forgotten customers arriving for service.

Answer B is incorrect. The service consultant should never invite customers to come in for service without entering the customer into the appointment log. This practice could lead to the problem of having too many vehicles show up for repair or service.

Answer C is incorrect. Neither Service Consultant is correct.

Answer D is correct. Neither Service Consultant is correct. All appointments should be entered into the appointment log in order to better manage shop workflow and prevent overscheduling customers arriving for service.

50. An upset customer calls and expresses frustration because he nearly had an accident due to the brake pedal going all the way to the floor. The shop had performed a complete brake job the day before. Service Consultant A recommends that the customer return to the service facility in two days for a recheck of the brake system. Service Consultant B offers to send a wrecker to pick up the vehicle immediately due to the potentially unsafe condition of the vehicle. Who is correct?

TASK C.4

A. A only
B. B only
C. Both A and B
D. Neither A nor B

Answer A is incorrect. The service consultant should not allow the vehicle to be driven any more before it is checked out thoroughly.

Answer B is correct. Only Service Consultant B is correct. Sending a wrecker is the right reaction to this situation due to the serious safety liability involved.

Answer C is incorrect. Only Service Consultant B is correct.

Answer D is incorrect. Service Consultant B is correct.

SECTION 7

Appendices

PREPARATION EXAM ANSWER SHEET FORMS

ANSWER SHEET

1. ______
2. ______
3. ______
4. ______
5. ______
6. ______
7. ______
8. ______
9. ______
10. ______
11. ______
12. ______
13. ______
14. ______
15. ______
16. ______
17. ______
18. ______
19. ______
20. ______
21. ______
22. ______
23. ______
24. ______
25. ______
26. ______
27. ______
28. ______
29. ______
30. ______
31. ______
32. ______
33. ______
34. ______
35. ______
36. ______
37. ______
38. ______
39. ______
40. ______
41. ______
42. ______
43. ______
44. ______
45. ______
46. ______
47. ______
48. ______
49. ______
50. ______

ANSWER SHEET

1. ____________
2. ____________
3. ____________
4. ____________
5. ____________
6. ____________
7. ____________
8. ____________
9. ____________
10. ____________
11. ____________
12. ____________
13. ____________
14. ____________
15. ____________
16. ____________
17. ____________
18. ____________
19. ____________
20. ____________
21. ____________
22. ____________
23. ____________
24. ____________
25. ____________
26. ____________
27. ____________
28. ____________
29. ____________
30. ____________
31. ____________
32. ____________
33. ____________
34. ____________
35. ____________
36. ____________
37. ____________
38. ____________
39. ____________
40. ____________
41. ____________
42. ____________
43. ____________
44. ____________
45. ____________
46. ____________
47. ____________
48. ____________
49. ____________
50. ____________

ANSWER SHEET

1. ______________
2. ______________
3. ______________
4. ______________
5. ______________
6. ______________
7. ______________
8. ______________
9. ______________
10. ______________
11. ______________
12. ______________
13. ______________
14. ______________
15. ______________
16. ______________
17. ______________
18. ______________
19. ______________
20. ______________
21. ______________
22. ______________
23. ______________
24. ______________
25. ______________
26. ______________
27. ______________
28. ______________
29. ______________
30. ______________
31. ______________
32. ______________
33. ______________
34. ______________
35. ______________
36. ______________
37. ______________
38. ______________
39. ______________
40. ______________
41. ______________
42. ______________
43. ______________
44. ______________
45. ______________
46. ______________
47. ______________
48. ______________
49. ______________
50. ______________

ANSWER SHEET

1. ______________
2. ______________
3. ______________
4. ______________
5. ______________
6. ______________
7. ______________
8. ______________
9. ______________
10. ______________
11. ______________
12. ______________
13. ______________
14. ______________
15. ______________
16. ______________
17. ______________
18. ______________
19. ______________
20. ______________
21. ______________
22. ______________
23. ______________
24. ______________
25. ______________
26. ______________
27. ______________
28. ______________
29. ______________
30. ______________
31. ______________
32. ______________
33. ______________
34. ______________
35. ______________
36. ______________
37. ______________
38. ______________
39. ______________
40. ______________
41. ______________
42. ______________
43. ______________
44. ______________
45. ______________
46. ______________
47. ______________
48. ______________
49. ______________
50. ______________

ANSWER SHEET

1. ______________
2. ______________
3. ______________
4. ______________
5. ______________
6. ______________
7. ______________
8. ______________
9. ______________
10. ______________
11. ______________
12. ______________
13. ______________
14. ______________
15. ______________
16. ______________
17. ______________
18. ______________
19. ______________
20. ______________
21. ______________
22. ______________
23. ______________
24. ______________
25. ______________
26. ______________
27. ______________
28. ______________
29. ______________
30. ______________
31. ______________
32. ______________
33. ______________
34. ______________
35. ______________
36. ______________
37. ______________
38. ______________
39. ______________
40. ______________
41. ______________
42. ______________
43. ______________
44. ______________
45. ______________
46. ______________
47. ______________
48. ______________
49. ______________
50. ______________

ANSWER SHEET

1. ______
2. ______
3. ______
4. ______
5. ______
6. ______
7. ______
8. ______
9. ______
10. ______
11. ______
12. ______
13. ______
14. ______
15. ______
16. ______
17. ______
18. ______
19. ______
20. ______
21. ______
22. ______
23. ______
24. ______
25. ______
26. ______
27. ______
28. ______
29. ______
30. ______
31. ______
32. ______
33. ______
34. ______
35. ______
36. ______
37. ______
38. ______
39. ______
40. ______
41. ______
42. ______
43. ______
44. ______
45. ______
46. ______
47. ______
48. ______
49. ______
50. ______

Glossary

Actuator A device that delivers motion in response to an electrical signal.

A/D Converter Abbreviation for analog-to-digital converter.

Additive A chemical addition intended to improve a certain characteristic of the material or fluid.

Aftercooler A charge air cooling device, usually water-cooled.

Air Compressor A engine-driven mechanism for supplying high pressure air to the truck brake system.

Air Filter A device that minimizes the possibility of impurities entering the intake system.

Altitude Compensation System An altitude barometric switch and solenoid used to provide better driveability at 1000 feet plus above sea level.

Ambient Temperature Temperature of the surrounding air. Normally, it is considered to be the temperature in the service area where testing is taking place.

Amp Abbreviation for ampere.

Ampere The unit for measuring electrical current.

Analog Signal A voltage signal that varies within a given range (from high to low, including all points in between).

Analog-to-Digital Converter (A/D converter) A device that converts analog voltage signals to a digital format; located in the ECM.

Analog Volt/Ohmmeter (AVOM) A test meter used for checking voltage and resistance. Analog meters should not be used on solid state circuits.

Antifreeze A mixture added to water to lower its freezing point.

Armature The rotating component of a: (1) starter or other motor; (2) generator.

Articulation Pivoting movement.

ASE Acronym for Automotive Service Excellence, a trademark of the National Institute for Automotive Service Excellence.

Atmospheric Pressure The weight of the air at sea level; 14.696 pounds per square inch (psi) or 101.33 kilopascals (kPa).

Axis of Rotation The center line around which a gear or part revolves.

Battery Terminal A tapered post or threaded studs on top of the battery case for connecting the cables.

Bimetallic Two dissimilar metals joined together that have different bending characteristics when subjected to changes of temperature.

Blade Fuse A type of fuse that has two flat male lugs for insertion in female box connectors.

Blower Fan A fan that pushes or blows air through a ventilation, heater, or air conditioning system.

Bobtailing A tractor running without a trailer.

Boss Outer race of a bearing.

Bottoming A condition that occurs when the teeth of one gear touch the lowest point between teeth of a mating gear.

British Thermal Unit (BTU) A measure of heat quantity equal to the amount of heat required to raise 1 pound of water 1° F.

BTU Acronym for British thermal unit.

CAA Acronym for Clean Air Act.

Cartridge Fuse A type of fuse having a strip of low melting point metal enclosed in a glass tube. If an excessive current flows through the circuit, the fuse element melts at the narrow portion, opening the circuit and preventing damage.

Caster The angle formed between the kingpin axis and a vertical axis as viewed from the side of the vehicle. Caster is considered positive when the top of the kingpin axis is behind the vertical axis.

Cavitation A condition caused by bubble collapse.

C-EGR Cooled exhaust gas recirculation.

CFC Acronym for chlorofluorocarbon.

Charging Circuit The alternator (or generator) and associated circuit used to keep the battery charged and power the vehicle electrical system when the engine is running.

Charging System A system consisting of the battery, alternator, voltage regulator, associated wiring, and the electrical loads of a vehicle. The purpose of the system is to recharge the battery whenever necessary and to provide the current required to power the electrical components.

Check-Valve A valve that allows air to flow in one direction only.

Climbing A gear problem caused by excessive wear in gears, bearings, and shafts whereby the gears move sufficiently apart to cause the apex of the teeth on one gear to climb over the apex of another gear.

Clutch A device for connecting and disconnecting the engine from the transmission.

COE Acronym for cab-over-engine.

Coefficient of Friction A measurement of the amount of friction developed between two objects in physical contact when one of the objects is drawn across the other.

Coil Springs Spring steel spirals.

Compression Applying pressure to a spring or fluid.

Compressor Mechanical device that increases pressure within a circuit.

Condensation The process by which gas (or vapor) changes to a liquid.

Conductor Any material that permits the electrical current to flow.

Coolant Liquid that circulates in an engine cooling system.

Coolant Heater A component used to aid engine starting and reduce the wear caused by cold starting.

Coolant Hydrometer A tester designed to measure coolant specific gravity and determine antifreeze protection.

Cooling System System for circulating coolant.

Crankcase The housing within which the crankshaft rotates.

Cranking Circuit The starter circuit, including battery, relay (solenoid), ignition switch, neutral start switch (on vehicles with automatic transmission), and cables and wires.

Cycling (1) On-off action of the air conditioner compressor; (2) Repeated electrical cycling that can cause the positive plate material to break away from its grids and fall into the sediment base of the battery case.

Dampen To slow or reduce oscillations or movement.

Dampened Discs Discs that have dampening springs incorporated into the disc hub. When engine torque is transmitted to the disc, the plate rotates on the hub, compressing the springs. This action absorbs the torsional vibration caused by today's low RPM, high torque, engines.

Data Links Circuits through which computers communicate with other electronic devices such as control panels, modules, sensors, or other computers.

Deburring To remove sharp edges from a cut.

Deflection Bending or moving to a new position as the result of an external force.

DER Acronym for Department of Environmental Resources.

Detergent Additive An additive that helps keep metal surfaces clean and prevents deposits. These additives suspend particles of carbon and oxidized oil in the oil.

Diagnostic Flow Chart A chart that provides a systematic approach to electrical system and component troubleshooting and repair. They are found in service manuals and are vehicle make and model specific.

Dial Caliper A measuring instrument capable of taking inside, outside, depth, and step measurements.

Digital Binary Signal A signal that has only two values: on and off.

Digital Volt/Ohmmeter (DVOM) A test meter recommended for use on solid state circuits.

Diode Semiconductor device formed by joining P-type semiconductor material with N-type semiconductor material. A diode allows current to flow in one direction, but not in the opposite direction.

DOT Acronym for Department of Transportation.

Driven Gear A gear that is driven by a drive gear, by a shaft, or by some other device.

Drive or Driving Gear A gear that drives another gear.

Driveline The propeller or driveshaft and universal joints that link the transmission output to the axle pinion gear shaft.

Driveline Angle The alignment of the transmission output shaft, driveshaft, and rear axle pinion centerline.

Drive Shaft Assembly of one or two universal joints connected to a shaft or tube; used to transmit torque from the transmission to the differential.

Drive Train An assembly that includes all torque transmitting components from the rear of the engine to the wheels.

ECM Acronym for electronic control module.

ECU Acronym for electronic control unit.

Eddy Current Circular current produced inside a metal core in the armature of a starter motor. Eddy currents produce heat and are reduced by using a laminated core.

Electricity The movement of electrons from one location to another.

Electromotive Force (EMF) The force that moves electrons between atoms. This force is the pressure that exists between the positive and negative points. This force is measured in units called volts that denote the charge differential.

Electronically Erasable Programmable Memory (EEPROM) Computer memory that enables write-to functions.

Electrons Negatively charged particles orbiting every nucleus.

EMF Acronym for electromotive force.

Engine Brake A hydraulically operated device that converts the engine into a power absorbing mechanism.

Environmental Protection Agency An agency of the United States government charged with the responsibilities of protecting the environment.

EPA Acronym for the Environmental Protection Agency.

Exhaust Brake A slide mechanism which restricts the exhaust flow, causing exhaust back pressure to build up in the engine's cylinders. The exhaust brake actually transforms the engine into a power absorbing air compressor driven by the wheels.

False Brinelling The polishing of a surface that is not damaged.

Fatigue Failures Progressive destruction of a shaft or gear teeth usually caused by overloading.

Fault Code A code that is recorded into the computer's memory.

Federal Motor Vehicle Safety Standard (FMVSS) A federal standard that specifies that all vehicles in the United States be assigned a Vehicle Identification Number (VIN).

Fixed Value Resistor An electrical device that is designed to have only one resistance rating, which should not change, for controlling voltage.

Flammable Any material that will easily catch fire or explode.

Flare To spread gradually outward in a bell shape.

Foot-Pound An English unit of measurement for torque. One foot-pound is the torque obtained by a force of one pound applied to a foot-long wrench handle.

Fretting A result of vibration that the bearing outer race can pick up the machining pattern.

Fuse Link A short length of smaller gauge wire installed in a conductor, usually close to the power source.

Fusible Link A term often used for an insulated fuse link.

Gear A disk-like wheel with external or internal teeth that serves to transmit or change motion.

Gear Pitch The number of teeth per given unit of pitch diameter, an important factor in gear design and operation.

Ground The negatively charged side of a circuit. A ground can be a wire, the negative side of the battery, or the vehicle chassis.

Grounded Circuit A shorted circuit that causes current to return to the battery before it has reached its intended destination.

Harness and Harness Connectors The vehicle's electrical system wiring that provides a convenient starting point for tracking and testing circuits.

Hazardous Materials Any substance that is flammable, explosive, or is known to produce adverse health effects to people or the environment.

Heads-Up Display (HUD) A technology used in some vehicles that superimposes data on the driver's normal field of vision. The operator can view the information, which appears to "float" just above the hood at a range near the front of a conventional tractor or truck. This allows the driver to monitor conditions such as road speed without interrupting his view of traffic.

Heater Control Valve A valve that controls the flow of coolant into the heater core from the engine.

Heat Exchanger A device used to transfer heat, such as a radiator or condenser.

Heavy-Duty Truck A truck that has a gross vehicle weight (GVW) of 26,001 pounds or more.

High-Resistant Circuits Circuits that have an increase in circuit resistance, with a corresponding decrease in current.

High-Strength Steel A low alloy steel that is stronger than hot-rolled or cold-rolled sheet steels.

Hinged Pawl Switch The simplest type of switch; one that makes or breaks the current of a single conductor.

HUD Acronym for heads-up display.

Hydrometer A tester designed to measure the specific gravity of a liquid.

Inboard Toward the centerline of the vehicle.

In-Line Fuse A fuse that is in series with the circuit in a small plastic fuse holder, not in the fuse box or panel. It is used, when necessary, as a protection device for a portion of the circuit even though the entire circuit may be protected by a fuse in the fuse box or panel.

Installation Templates Drawings supplied by some vehicle manufacturers to allow the technician to correctly install the accessory. The templates available can be used to check clearances or to ease installation.

Insulator A material, such as rubber or glass, that offers high resistance to electron flow.

Integrated Circuit A component containing diodes, transistors, resistors, capacitors, and other electronic components mounted on a single piece of material and capable to perform numerous functions.

Jacob's Engine Brake An engine brake, named for its inventor. A hydraulically operated device that converts a power-producing diesel engine into a power-absorbing retarder.

Jumper Wire A wire used to temporarily by-pass a circuit or components for electrical testing. A jumper wire consists of a length of wire with an alligator clip at each end.

Jump Start The procedure used when it becomes necessary to use a boost battery to start a vehicle with a discharged battery.

Kinetic Energy Energy in motion.

Lateral Runout The wobble or side-to-side movement of a rotating wheel.

Laser Beam Alignment System A two- or four-wheel alignment system using wheel-mounted instruments to project a laser beam to measure toe, caster, and camber.

Linkage A system of rods and levers used to transmit motion or force.

Low-Maintenance Battery A conventionally vented, lead/acid battery, requiring normal periodic maintenance.

Magnetorque An electromagnetic clutch.

Maintenance-Free Battery A battery that does not require the addition of water during normal service life.

Maintenance Manual A publication containing routine maintenance procedures and intervals for vehicle components and systems.

NATEF Acronym for National Automotive Education Foundation.

National Automotive Technicians Education Foundation (NATEF) A foundation with a program for certifying secondary and post-secondary automotive and heavy-duty truck training programs.

National Institute for Automotive Service Excellence (ASE) A nonprofit organization that has an established certification program for automotive, truck, auto body repair, engine machine shop technicians, and parts specialists.

NHTSA Acronym for National Highway Traffic Safety Administration.

NIOSH Acronym for National Institute for Occupational Safety and Health.

NLGI Acronym for National Lubricating Grease Institute.

NOP Acronym for nozzle opening pressure. Pressure in an injector nozzle opens at inoperation. Also known as valve opening pressure (VOP).

OEM Acronym for original equipment manufacturer.

Off-Road With reference to unpaved, rough, or ungraded terrain on which a vehicle will operate. Any terrain not considered part of the highway system falls into this category.

Ohm A unit of electrical resistance.

Ohm's Law Basic law of electricity stating that in any electrical circuit, current, resistance, and pressure work together in a mathematical relationship.

On-Road With reference to paved or smooth-graded surface on which a vehicle will operate; part of the public highway system.

Open Circuit An electrical circuit whose path has been interrupted or broken, either accidentally (a broken wire) or intentionally (a switch turned off).

Oscillation Movement in either fore/aft or side-to-side direction about a pivot point.

OSHA Acronym for Occupational Safety and Health Administration.

Output Driver Electronic switch that the computer uses to control the output circuit. Output drivers are located in the output ECM.

Oval Condition that occurs when a tube is egg-shaped.

Overrunning Clutch A clutch mechanism that transmits power in one direction only.

Overspeed Governor A governor that shuts off fuel at a specific RPM.

Oxidation Inhibitor An additive used with lubricating oils to keep oil from oxidizing at high temperatures.

Parallel Circuit An electrical circuit that provides two or more paths for current flow.

Parallel Joint Type A type of drive shaft installation whereby all companion flanges and/or yokes in the complete driveline are parallel to each other, with the working angles of the joints of a given shaft being equal and opposite.

Parking Brake A mechanically applied brake used to prevent a parked vehicle's movement.

Parts Requisition A form that is used to order new parts on which the technician writes the part(s) needed along with the vehicle's VIN.

Payload The weight of the cargo carried by a truck, not including the weight of the body.

Pitting Surface irregularities resulting from corrosion.

Polarity The state of charge differential, either positive or negative.

Pole The number of input circuits made by an electrical switch.

Pounds per Square Inch (psi) A unit of English measure for pressure.

Power A measure of work being done factored with time.

Power Flow The flow of power from the input shaft through one or more sets of gears.

Power Train The collection of components that transfers energy from the engine to the wheels in a vehicle.

Pressure The force applied to a definite area measured in pounds per square inch (psi) English or kilopascals (kPa) metric.

Pressure Differential The difference in pressure between any two points of a system or a component.

Printed Circuit Board Electronic circuit board made of thin, nonconductive material onto which conductive metal has been deposited. The metal is then etched by acid, leaving lines that form conductors for the circuits on the board. A printed circuit board can hold many complex circuits.

Programmable Read-Only Memory (PROM) Electronic component that contains program information specific to vehicle model calibrations.

PROM Acronym for Programmable Read-Only Memory.

psi Acronym for pounds per square inch.

P-type Semiconductors Positively biased semiconductors.

RAM Acronym for random access memory; the main memory of a computing device.

Ram Air Air forced into the engine housing or passenger compartment by the forward motion of the vehicle.

Random Access Memory (RAM) Memory used during computer operation to store temporary information. The microcomputer can write, read, and erase information from RAM, which is electronically retained.

RCRA Acronym for Resource Conservation and Recovery Act.

Reactivity The characteristic of a material that enables it to react violently with air, heat, water, or other materials.

Read-Only Memory (ROM) Memory used in microcomputers to store information permanently.

Recall Bulletin A bulletin that pertains to special situations that involve service work or replacement of components in connection with a recall notice.

Reference Voltage The voltage supplied to a sensor by the computer, which acts as base line voltage; modified by the sensor to act as an input signal; usually 5 volts direct current (VDC).

Relay An electric switch that allows a small current to control a much larger one. It consists of a control circuit and a power circuit.

Reserve Capacity Rating The ability of a battery to sustain a minimum vehicle electrical load in the event of a charging system failure.

Resistance Opposition to current flow in an electrical circuit.

Resource Conservation and Recovery Act (RCRA) Law that states that after using hazardous material, it must be properly stored until an approved hazardous waste hauler arrives to take it to a disposal site.

Revolutions per Minute (rpm) The number of complete turns a shaft makes in one minute.

Right to Know Law A law passed by the federal government and administered by the Occupational Safety and Health Administration (OSHA) that requires any company that uses or produces hazardous chemicals or substances to inform its employees, customers, and vendors of any potential hazards that may exist in the workplace as a result of using the products.

Ring Gear The gear around the edge of a flywheel.

ROM Acronym for read-only memory.

Rotary Oil Flow A condition caused by the centrifugal force applied to the fluid as the converter rotates around its axis.

Rotation A term used to describe a gear, shaft, or other device when it is turning.

Rotor Rotating part of the alternator that provides the magnetic fields necessary to create a current flow; the rotating member of an assembly.

RPM or rpm Acronym for revolutions per minute.

Runout Deviation or wobble of a shaft or wheel as it rotates, which is measured with a dial indicator.

Semiconductor Solid state material used in diodes and transistors.

Sensing Voltage The voltage that allows the regulator to sense and monitor the battery voltage level.

Sensor An electronic device used to monitor conditions for computer control requirements. An input circuit device.

Series Circuit A circuit connected to a voltage source with only one path for electron flow.

Series/Parallel Circuit Circuit designed so that both series and parallel conditions exist within the same circuit.

Service Bulletin Publication that provides the latest service tips, field repairs, product improvements, and related information of benefit to service personnel.

Service Contract A type of extended warranty that many customers purchase when they buy their vehicle. These contracts are basically an insurance policy that pays for the repairs that are listed in the policy.

Service Manual A manual, published by the manufacturer, that contains service and repair information for all vehicle systems and components.

Short Circuit An undesirable connection between two worn or damaged wires. The short occurs when the insulation is worn between two adjacent wires and the metal in each wire contacts the other or when wires are damaged or pinched.

Single-Axle Suspension A suspension with one axle.

Single Reduction Axle Any axle assembly that employs only one gear reduction through its differential carrier assembly.

Solenoid An electromagnet that is used to conduct electrical energy in mechanical movement.

Solid Wires Single-strand conductor.

Solvent Substance which dissolves other substances.

Spade Fuse Term used for blade fuse.

Spalling Surface fatigue that occurs when chips, scales, or flakes of metal break off.

Specialty Service Shop A shop that specializes in areas such as engine rebuilding, transmission/axle overhauling, brake, air conditioning/heating repairs, and electrical/electronic work.

Specific Gravity The ratio of a liquid's mass to that of an equal volume of distilled water.

Spontaneous Combustion A reaction in which a combustible material self-ignites.

Stall Test Test performed when there is a malfunction in the vehicle's power package (engine and transmission) to determine which of the components is at fault.

Starter Circuit The circuit that carries the high current flow and supplies power for engine cranking.

Starter Motor Device that converts electrical energy from the battery into mechanical energy for cranking.

Starter Safety Switch Switch that prevents vehicles with automatic transmissions from being started in gear.

Static Balance Balance at rest or still balance.

Stepped Resistor A resistor designed to have two or more fixed values, available by connecting wires to one of several taps.

Storage Battery A battery that provides a source of direct current electricity for both the electrical and electronic systems.

Stranded Wire Wire that is made up of a number of small solid wires, generally twisted together, to form a single conductor.

Sulfation Condition that occurs when sulfate is allowed to remain on the battery plates for a long time, causing two problems: (1) It lowers the specific gravity levels, increasing the danger of freezing at low temperatures; (2) in cold weather a sulfated battery may not have the reserve power needed to crank the engine.

Swage To reduce or taper.

Switch Device used to control on/off and direct the flow of current in a circuit. A switch can be under the control of the driver or can be self-operating through a condition of the circuit, the vehicle, or the environment.

Tachometer Instrument that indicates shaft rotating speeds.

Technical Service Bulletin A document produced by the manufacturer to assist technicians in fixing pattern failures.

Throw (1) Offset of a crankshaft; (2) Number of output circuits of a switch.

Time Guide Used for computing compensation payable by the truck manufacturer for repairs or service work to vehicles under warranty.

Timing The phasing of events to produce a desired action, such as ignition.

Torque Twisting force.

Torque Converter A device, similar to a fluid coupling, that transfers engine torque to the transmission input shaft and can multiply engine torque.

Toxicity A statement of how poisonous a substance is.

Tractor A motor vehicle that has a fifth wheel and is used for pulling a semitrailer.

Transistor Electronic device produced by joining three sections of semiconductor materials; used as a switching device.

Tree Diagnosis Chart Chart used to provide a logical sequence for what should be inspected or tested when troubleshooting a repair problem.

Vacuum Pressure values below atmospheric pressure.

Vehicle Retarder An engine or driveline brake.

VIN Acronym for vehicle identification number.

Viscosity Resistance to flow or fluid sheer.

Volt The unit of electromotive force.

Voltage-Generating Sensors Devices which produce their own voltage signal.

Voltage Limiter Device that provides protection by limiting voltage to the instrument panel gauges to approximately 5 volts.

Voltage Regulator Device that controls the current produced by the alternator and thus the voltage level in the charging circuit.

VOP Acronym for valve opening pressure, used by Caterpillar to denote nozzle opening pressure (NOP).

Watt Measure of electrical power.

Watt's Law A law of electricity used to calculate the power consumed in an electrical circuit, expressed in watts. It states that power equals voltage multiplied by current.

Windings (1) The three bundles of wires in the stator; (2) coil of wire in a relay or similar device.

Work (1) Forcing a current through a resistance; (2) the product of a force.

Yield Strength The highest stress a material can stand without permanent deformation or damage, expressed in pounds per square inch (psi).

Notes

Notes

Notes

Notes

Notes

Notes